Grow, Freeze and Cook
YOUR GARDEN ON A PLATE

Grow, Freeze and Cook

YOUR GARDEN ON A PLATE

Sheila Howarth

Pelham Books - London

First published in Great Britain by PELHAM BOOKS LTD
52 Bedford Square, London WC1B 3EF
FIRST PUBLISHED JULY 1974
SECOND IMPRESSION DECEMBER 1974

ISBN 0 7207 0675 0

Printed in Great Britain by
Hollen Street Press Limited at Slough, Berkshire
and bound by James Burn at Esher, Surrey.

CONTENTS

LIST OF ILLUSTRATIONS

All photographs are by Robert Corbin

INTRODUCTION

The greengrocer's shop in large, and even small towns is losing all signs of the seasons. Imported fruit and vegetables make it possible to have your particular favourites all through the year – at a price.

They may look, and be called 'fresh'. But only by a determined stretch of imagination and lapse of taste-buds, could their flavour be described as fresh.

You would not expect it anyway of tree fruit and root crops, grown in another climate, kept in cold store for months, then released over here when home grown supplies are dwindling. It should not be expected either of perishable varieties.

You can get raw (meaning uncooked, rather than fresh) luxury fruit and vegetables, such as asparagus, zucchini, French beans, apricots, plums, strawberries and other soft fruit, at any time.

The excitement of an out of season treat may be enough in itself, but the flavour will bear little resemblance to the same thing harvested at the peak of perfection moment, and eaten or frozen immediately. Until you have grown your own produce you will never know what really fresh fruit and vegetables taste like.

I am still surprised at the number of well-travelled and seemingly discerning eaters I have entertained, who have either never met or heard of the vegetable or fruit I serve, or have met them but never known their true taste. All are raised in my own garden (which takes a back seat in a busy life), with no greenhouse or cloches (they take too much time and nannying) and an invaluable part-time slave-driver.

9

He spends much of the time mowing and fiddling with the four acres which do not produce things to eat.

The shop which used to have the celery and carrots still crisp and swaddled in the soil that grew them, has become a department of a multiple store. You help yourself to a hygienically sealed pack of washed, weighed, and seemingly polished vegetables, which have had all the trouble and taste taken from them. Also most of the goodness.

This state of affairs agitated by rising food costs has happily resulted in a surge of home-growing, particularly by newly married couples with their first home, first garden and first child. Not only the economics but the wish to control the chemicals and fertilisers, which inadvertently may become part of our diet, has started the rush. The beguiling but much mis-used and mis-interpreted beckoning label 'organically grown', has also helped the back to nature trend – to the delight of a latent army of pests and diseases!

The result is that there is now a waiting list for allotments in a great many urban districts.

One of the great advantages of a freezer is that you can choose varieties which will be ready to pick when you know you will be there and not on holiday. Any thrifty gardener used to dread a glut. It meant giving away the surplus – a time-consuming and not always appreciated form of charity; now charity begins and stays at home. Also the family never gets satiated with one particular crop when it is in season. Even asparagus eaten every day can pall.

With crops which keep fit and in their prime for a great length of time, such as Brussels sprouts, you can choose just the right day to freeze a large batch, when the ground does not stick to your feet, and you have the time to cope. With quick growing crops like summer spinach, which go to seed almost overnight, you can rescue them before they are in danger.

With a freezer, you are boss of both garden and kitchen. Never again need you say 'Sorry its beans again, but they need eating.'

Part One

THE BEGINNING AND THE REASON WHY

Flavour: Variety: quality: convenience: economy. These are THE BENEFITS of growing and freezing your own crops. The order of merit in which you place these advantages will naturally depend on your style of life and eating. For the unadventurous who cannot tell one variety of pea or bean from another, it may be entirely economic. For the gourmet it opens up new worlds at a fraction of the cost of the shop-bought equivalent.

FLAVOUR
Vegetables and fruit grown for the wholesale market have to pass an entirely different fitness test from those grown in a garden. The crops which get to the shops have been chosen for their resistance to pests and disease, evenness and speed of growth, eye-appeal, and their ability to travel. This obviously requires a tough constitution. They have to run the gauntlet of mass handling – picking, grading, packaging, transport distribution – before they reach your shopping bag. Unfortunately many of the finest flavoured varieties are too delicate to stand the strain of such a marathon, but are just as easy to grow in the garden as market types.

Freshness is an integral part of flavour, and only when you have the water boiling before picking the sweet corn or shelling newly-picked peas, will you ever know what they really taste like, or how little cooking they need.

VARIETY
Few people realise what a tremendous range of different types

of vegetables and fruits they can choose from, because they don't know they exist. You find them on private dinner tables, but only rarely in shops, and at a rare old price. Some of the crops may sound exotic, but all can be grown easily by the beginner without a greenhouse.

Some soft fruits are impossible to buy, except perhaps at a village shop, because they will not keep, some not even overnight. They are also too difficult to pick to be of any financial interest to the grower. You must be prepared to grow them with love, knowing the reward will be in the eating.

Many of the unusual have vanished from the popular general seed catalogues, because only country grandparents remember when to harvest and how to cook them.

QUALITY

The home grower can choose vegetables and fruit entirely for their eating value. He does not have to worry about getting the maximum tonnage per acre, or whether the crops will mature all together at the drop of a hat, as it were, so machines can be sent in and 'clean-up' in a day.

With your own crops right under your nose you can pick the best for the day and for the freezer, and at the same time afford to discard anything past its prime, or turn it into purée for soups or ices.

CONVENIENCE

A tin of baked beans may sound easier if you are near shops, but you cannot lay hands on them at all hours of the day and holidays, even if this is the limit of your gastronomic desires.

With your own produce within walking distance (garage, tool shed, spare bedroom or what-have-you); there is no worry about being marooned by strikes, punctures or even a broken leg.

The most distressing of all calamities for hospitable families is also dispelled – week-end droppers-in.

There is none of the rushed buying frenzy before bank holidays, trying to guess how much of what might be needed and how long it will keep fresh.

You have only to compare the prices of ready frozen fruit and vegetables, and fresh shop-bought ones, with those you grow yourself, to realise what a ludicrous discrepancy exists. Often the shop-bought produce, particularly peas and beans, are inferior to the mass marketed frozen ones, because by the time they reach your table they are several days 'fresh'. Whereas the giants of the freezing industry send their freezing units out to the crops so they are in the bag in a pristine state within an hour. But they are field peas, rather than garden peas, and bear no taste relation to the varieties you can grow yourself.

For the price of a few packets of seed, a little fertiliser and as little or as much time as you want to give, you can have both plain and fancy crops to eat fresh all through the year.

You can also economise on birthdays and Christmas presents. Assorted packets of loganberries, yellow raspberries and black currants; red broad beans, zucchini blue coco French beans, are welcomed with far more enthusiasm than a potted plant or bottle of wine.

PLACE, PLANTS AND PROCEDURE

No two gardens are alike, and the amount of space you are prepared to give to crops must be an individual choice. A small corner garden in a side road of a market town which I have passed regularly for many years, is obviously manned and planned by an individualist. The neighbouring 'semis' have the uniform rectangle of grass with a specimen shrub in the middle; narrow borders of bedding plants and a clematis over the porch. This garden grows nothing but vegetables, in proud rows, fully exposed to public view behind a low wall, crossways, sideways and up to the front door. I have seen it at all seasons and never caught it napping or messy.

This, to many, would seem going to extremes for the sake of the stomach, but there is no need to think of vegetable patches as inevitable eyesores. Hide them away behind the garage, or a fence at the bottom of the garden if you must, but there is no reason why edible crops should not be combined with shrubs and flowers.

The chief reason for keeping crops to themselves in a kitchen garden is convenience, and crop rotation. They need sun, a rich well-dug soil, and regular weeding which is easier if they are in well-spaced rows.

Permanent crops like asparagus or perpetual strawberries are not a pretty sight for much of the year, particularly when the strawberries have to be netted, but other perennials such as the spectacular globe artichokes, or perennial herbaceous herbs like tarragon, fennel, balm, chives, marjoram and sorrel make splendid border plants, though they all die out of sight in winter.

Whatever the size of your garden and wherever you decide to grow your fruit and vegetables there are certain common sense points to remember.

They need air, light and sun, so don't tuck them away in a dark corner. Neither will they stand much chance in a wind-tunnel.

Grow them within easy reach of water – standpipe, water-butt or hose. Easy reach of hard paths for the sake of the wheelbarrow and your shoes.

With a freezer it is not so important to have the vegetable garden as close to the kitchen as possible in winter. You will have harvested everything except some greens before the bad weather starts, and can keep clear of it entirely unless the mood and weather suit you.

WHERE TO GROW

UNLIKELY PLACES FOR LIKELY CROPS

Containers: Large and small decorative or humdrum boxes, tin cans, tubs, urns, even hanging baskets, can be used, within reason, for anything which grows in the open ground. They must, though, have enough depth of soil for their various root sizes. Window boxes are excellent for annual or small herbs and radishes. Dwarf fruit trees are happy in tubs. Strawberries will grow in special barrels resembling colanders, punctured with planting holes. Bush marrows and squashes keep their dignity when you constantly take their fruit, and simply produce more. They take kindly to tub and patio life, where you can give them all the moisture and feeding they thrive on. This goes for all container crops whether in coal scuttles or wellington boots. The position and soil must be right and they should never dry out.

Windscreens: Jerusalem artichokes grow to 5–6 ft, and will happily protect you, or more tender crops. Sweet corn serves the same purpose, but needs to be planted in a block rather than a row and strong winds can topple them over as they have shallow roots rather than deep tubers like the artichokes to anchor them.

Edgings: Alpine strawberries, parsley, shallots, garlic, early carrots and annual herbs such as basil, chervil and dill.

Boundary Fences and Eyesores: Trailing and climbing vegetables such as beans, cucumbers or marrows can be most attractive, particularly the blue podded variety of bean. The self-climbers such as runner beans will need to be kept in hand and the trailers given some support to keep them in position. One of the quickest and most useful ways of disguising a new garage, shed, oil tank or anything which offends is to plant a cultivated blackberry on the sunniest side. They, too, make excellent boundary hedges against wires or a fence. The trailing raspberry Zeva can be used in the same way but is less vigorous and may need protection from birds.

Flower Borders: Bush tomatoes can be dotted among flowers in open beds or against walls. The small fruited varieties are most decorative and children eat them by the handful. Rhubarb beet is another edible growing ornament, and if you have space for some globe artichokes at the back of a border, you can enjoy their wonderfully cut grey furry leaves throughout the year, as well as removing their 'globes' for eating before the 'chokes' become too large.

17

This will be governed almost entirely by the space you have in the garden, in the freezer, economy and personal taste. You can concentrate on family favourites – peas, beans, raspberries, strawberries – or go in for smaller quantities of a much larger variety which are not easy to buy locally and are always expensive if you can.

There is little point in growing things which you eat only occasionally and are always available in shops, like cabbage, the broccolis and celery. You won't mind the extra cost instead of the gardening when you feel like them. But if these are frequently on the family menu, they will save a great deal of money and time to grow.

It seems a waste of space to freeze root crops which are perfectly happy to be stored, but by growing early varieties of such plants as carrots and beetroot, you score on several counts. Unlike maincrop varieties they do not store well but they are sweeter and mature quickly, you have them out of the ground before they get riddled with carrot fly, and you can use the space for something else at once. It also eliminates the horrible task of plunging a hand into a container of sand, groping for hard and not rotten roots.

You can also 'catch' and freeze other early types of root crops while they are in their baby-prime and not woody. Beetroot and turnips for example. They take only a fraction of the cooking time of the 'keepers' and taste remarkably better.

Some fruit and vegetables freeze better than others, but far more will take the treatment than commercial lists suggest (they have to please a massive average eater and cook by mass production).

The greatest joy of a freezer is that we can have unusual fresh home-grown produce all through the year. It has taken much experimenting over many years in both growing and freezing methods, some of which I readily admit were failures which you will be warned against.

In the following list I have given named varieties only when I have found them outstandingly better than others in flavour and texture, whether eaten fresh before or after freezing.

All can be grown without heat in most areas, but if you are in a particularly cold one, it is best not to embark on things like sweet corn unless you can start them off on a window ledge or in a cold frame, or they will not have time to ripen in the open.

Some of the vegetables and herbs can be bought from garden centres and shops, or ordered from the general lists of any leading seedsmen. I have chosen mine from the catalogues of Thompson & Morgan, Ipswich, Suffolk; Unwins, Histon, Cambridge; and Samuel Dobie & Sons, Llangollen, Denbighshire. They all include the more unusual varieties and the first two recommend those which are specially good for freezing.

VEGETABLES
Pods and Fruits
Artichoke Globe
Asparagus pea
Beans. *Broad Bean*. Masterpiece, Red Epicure, Gillette New
    Imperial Longpod
    French climbing. Violet podded French stringless, Royalty
Romano
    French dwarf. Remus, Cherokee Wax Pod, Tendergreen
    Runner. Goliath
Cucumber. Ridge, Burpless, Burlee Hybrid
Marrow and Squash. Cocozelle, Gold Nugget, Zucchini,
    Courgette
Peas. Green Recette, Hurst's Green Shaft, Victory Freezer
    Mange Tout. Dwarf Sugar, Dwarf de Grace
    Petit Pois. Gullivert, Cobri
Sweet Corn. Early Xtra, Kelvedon Glory
Tomato. Gardeners Delight. Fruit (peels without scalding),
    Yellow Perfection
Leaves, Shoots and Stems
Asparagus
Brussels sprouts. British Allrounder, Focus, King Arthur,
    Rubine-red, Prince Astrol

Broccoli. Purple sprouting, white sprouting
Cabbage
Calabrese. Green sprouting
Cauliflower
Celery. Greensnap
Chicory. Witloof or Brussels
Chinese Cabbage
Good King Henry. Chenopodium bonus-henricus
Kale. Green Curled, Dwarf and Tall
Kohl Rabi
Leeks
Lettuce. Little Gem
Seakale Beet (Swiss Chard). Vintage Green, Ruby Chard
Spinach Beet. Perpetual or Winter
Spinach. Monach Long Standing, Cleanleaf, New Zealand
Roots and Bulbs
Artichoke. Chinese
    Jerusalem
Beetroot. Gold, Snow White, Ruby Queen (use top too)
Carrot. Sweetheart
Celeriac
Celtuce
Garlic
Hamburg Parsley
Onions
Parsnips
Potato. Aura
Salsify and scorzonera
Shallot
Turnip

HERBS
The ones worth freezing are the tender half hardy ones which
die in the autumn, whose pungency can never be captured
in a dry state. Also some of the herbaceous perennial ones
which come up each year but are out of sight in winter
when we may need them.
Basil
Chervil

Chives
Dill
Fennel
Mint
Purslane
Sorrel
Summer Savory
Sweet marjoram
Tarragon

FRUIT
Personal favourites listed but any variety possible.
Apples
Apricots
Blackberry. Oregon Thornless
Blueberry. (Improved Hybrid variety of wild bilberry)
Boysenberry. Thornless
Cherry, Dessert and cooking
Currants. Black, white, red
Damson
Fig
Gooseberry. red, green, and white
Japanese Wineberry
Loganberry, thornless
Nectarine and peach
Pear
Plums
Raspberry. Zeva perpetual fruiting, yellow and red
Rhubarb
Strawberry. Gento perpetual fruiting, Alpine Baron
    Solemacher
Youngberry, thornless

RANDOM HARVEST
Bilberries – moorland
Blackberries – hedges, scrubland etc
Cockles – Shore and esturies at low tide
Elderflower and berries
Mushrooms and edible fungi – woods, fields etc

Mussels – Rocks at low tide – in sea bed
Samphire – Salty marshes

HOW TO GROW

THOUGHTS BEFORE YOU LEAP

You can forget all about that head-throbbing instruction 'sow at fortnightly intervals for succession' if you and your freezer can cope with generous loads at a time. Choose varieties which mature just when YOU want them to. It will not mean that you have to deal with the whole crop in a day. Unlike field and commercial types, garden crops, even the hybrids which boast uniformity, will still mature unevenly.

Don't grow and freeze so much of one thing that you are still eating the frozen ones when the next year's crop is coming up in the garden. The main purpose of a deep freeze is to have home-grown fresh fruit and vegetables OUT of season!

If you prefer quick growing summer vegetables to long standing greens and root crops, give most, or all space to them, and buy the others as you want them. You will be able to dig over the whole plot in early autumn, then forget about it until the following spring, when it will need only raking down after the winter has 'weathered' it.

Don't be put off by illustrations of vegetables growing in strict soldier rows with not a weed in sight. If you are gowing for the table, via the freezer, rather than the eye and competitions, get your priorities in order. Spend more time picking the baby crops which turn from flower to over-mature vegetables and fruit almost overnight, than in weeding. You need particularly to keep a watch on fast growing crops like beans, peas and the squash family.

As long as the soil is fertile it matters little whether the bean seeds are planted with their eyes up, down or sideways – at regular inches apart or in a wayward scattered line or trench. The peas will be more likely to survive bird attacks if weeds growing actually among them are left alone. You will only disturb the shallow roots of the crop if you try to pull them out. The Irish grow the most splendid potatoes in hillsides of nettles. Many weeds and vegetables help each other, and if you have only little time and limited knowledge, you can do more harm than good with a hoe or chemical weedkiller. There is no reason in being a perfectionist except to satisfy yourself or an audience, and straight rows and weeds won't alter the taste of what you grow.

They will, though, diminish the performance of permanent crops and long standing ones – cane fruit, rhubarb, asparagus, globe artichokes and root vegetables. These cannot take quite such a slap-happy approach and need some regular sorting out to keep the weeds from competing for breathing space above and below the ground belt.

SOIL

These have for generations been described in colourful terms by gardeners wishing theirs was another kind, but can generally be classed as either light or heavy, early or late, hungry or fertile. Within these general descriptions there are niceties of variation such as clay, loam, medium loam, loamy sand, stony, peaty and chalky soil.

None of them is likely to be perfect. Happily the draw-backs of one are compensated for by certain advantages which others lack. A heavy clay soil is a back-breaker to cultivate but is extremely fertile, does not dry out in summer and grows magnificent crops with the minimum attention of watering and hoeing. On the other hand a sandy soil can be dug on almost any day in the year by the frailest of us not actually bed or house bound. The drawback here is that plants begin to keel over after a week of hot sunny rainless days – usually when there is a restriction on the use of a garden hose.

Make the most of what is good about your soil, and try to lessen the drawbacks.

SOIL TYPES

Clayey: Heavy Loam, Clay loam, Clay.
GOOD POINTS. Well supplied with plant food.
BAD POINTS. Too cold for early crops. Heavy to work

when wet. Can waterlog in winter. Sticks to feet. Cakes hard and cracks in dry weather.

CURE. Dig thoroughly in autumn to expose soil lumps to winter frosts. Incorporate coarse humus making stuff such as compost, peat, strawy stable manure if possible. These stop the fine clay particles from compacting, and build up a crumb structure.

Stony

GOOD POINTS. Usually free draining. Can be worked early in the season.

BAD POINTS. Difficult to cultivate and dries out quickly in summer.

CURE. Rake off larger surface stones. Add plenty of manure, compost or fertiliser close to the surface. This will stop the goodness in the soil from draining through the stones too fast. Never dig deeply.

Peaty

GOOD POINTS. Easy to work. Fertile when limed and drained. Good for Azaleas, Rhododendrons, Heaths and Blueberry fruit.

BAD POINTS. Too acid for most plants, particularly vegetables.

CURE. Good drainage, Generous liming. Add loamy topsoil.

Chalky

GOOD POINTS. Best soil for several flowers and shrubs, especially rock plants.

BAD POINTS. Sticky and soft in wet weather.

CURE. These are hungry soils. Must always be only shallow-dug with plenty of humus-making materials and fertilisers.

Sandy: Loamy sand, Light sand.

GOOD POINTS. Warm. The best for early crops. Easy to eork even when wet. Free-draining because of open texture.

BAD POINTS. Hungry. Always ravenous for plant foods. Dries out very quickly and needs frequent watering during summer.

CURE. Improve matters with sticky manure such as cow and pig where possible. Fertilisers are essential and should not be dug in deeply. This soil is renowned as giving the least backache and most heartache.

Loam

GOOD POINTS. The perfect garden soil. It has all the advantages and none of the disadvantages of the others.

BAD POINTS. None except that you might take it for granted and give it no attention at all. In which case . . .

CURE. Keep it in its naturally pristine state with regular light dressings of lime, humus-makers and fertilisers. Dig over the ground in autumn.

There is no cause for panic or giving up the idea of growing your own crops, if the recommended manures and cures are not within cartage sight or the depth of your pocket. It depends on the size of your garden and where you live. Animal manure, rotted garden refuse, mushroom compost, peat, sewage sludge, spent hops and sugar beet factory waste are of tremendous value in vegetable growing because they feed the soil as well as preserving the structure. Even though, when they can be had locally in quantity, they are rarely enough in themselves and need to be supplemented by well-blended compound fertilisers which will supply nitrogen, phosphorus and potash – the elements most likely to need replacing in a heavily cropped vegetable garden.

Garden supply shops now bulge with compressed products in plastic bags which, though more expensive are easier to use and carry. Read the label and make sure you get the right one to cure what is wrong, or the right tonic for your particular soil.

Most vegetable crops grow perfectly happily without deep digging, as long as the topsoil is fertile and the subsoil not a bog. That dread term double-digging is really only a precaution when preparing a site on heavy soil for globe artichokes, asparagus and other crops which will occupy the ground for more than two years. It gives the opportunity to add bulky organic manure which rots down to provide humus.

At the same time you can dig out deep-rooted perennial weeds, docks, dandelions, thistles, bellbind, couch grass, etc., which would wreck the performance of any permanent crops. They can be destroyed by total weedkillers, but this means leaving the ground unplanted for up to a year (depending on what you use). Unharmful selective weedkilling chemicals including paraquat, sold as Weedol, can be used on annual weeds among crops. But the toughies simply shake their heads at these mild attacks, go to ground – slightly scorched – and emerge refreshed in another spot. All the time their powerful roots are robbing the goodness out of what you hope to eat.

Digging is the best way to combat permanent weeds where you want permanent plants. Beware of rotavating such nasties as couch and bellbind. They will be chopped into a multitude of potential offspring. Those left on the surface will wither but the buried ones can each turn into individual plants.

ROTATION

Though not absolutely essential it is better for both crops and soil if there is some rotation in the sequence of vegetables, so that the same kinds are not grown in exactly the same place every year. In large gardens it is possible to work on a three year plan. One part is devoted to peas, beans salads, onions and cauliflowers which do best on ground that has been manured. A second part will grow root crops such as carrots, parsnips and beetroot which fork or 'fang' if given fresh manure, and are better grown on land that was well manured the year before. The third plot is reserved for brassicas (members of the cabbage family) which suffer from a disease called 'club root' if grown for years in the same place. This disease is less troublesome if the soil is well-limed the season prior to planting.

Potatoes can be grown beside the root crops, and indeed will grow almost anywhere but, for really good crops they need well-manured (old) ground. In succeeding years the groups play musical chairs – the peas and beans going to the previous potato and root crop plot; the root crops replacing the brassicas; and the brassicas taking over the pea and bean soil.

A small garden cannot go in for such niceties, particularly if the edible crops are to be grown among flowers, bulbs and shrubs (ornamental plants). All you need to remember is the general rule of moving things around and knowing in advance where each crop is going, so that the appropriate diet can go into the ground beforehand. Vegetables, with the exception of the permanent crops, are like gipsies and need to keep on the move.

Rotating the crops is in the end an economy as it prevents a build up of pests and diseases in the soil. You don't have to waste time and money spraying them or feeding the inedible results to the compost heap.

If your garden is too small for potatoes or many root crops, it would be best to settle for a two-year rotation. One group consisting mainly of brassicas, the other being a mixture of anything else you fancy. Keep in mind, when moving them around, that if a deep-rooting crop is grown in the same place for several seasons the soil will become exhausted; and if a shallow-rooting crop is grown repeatedly it is not being used to its full depth.

Never trust your memory. Keep a garden book with a simple plan of what went where, each year. Once a crop has been cleared, it is remarkable how quickly you can forget it. Always try to plan crops so that dwarf and tall are interspersed enabling light and air to reach them all, without having to filter through dense forests.

VEGETABLES : Their Cultivation

ARTICHOKE : Chinese
One of the most curious and rare vegetables in this country, though not difficult to grow. It has odd ribbed tubers, ivory white, rather like cricket bails, which form in the soil up to 3 ins long and as much as an inch thick in the middle. Though called artichokes they belong to the dead nettle family Stachys tuberifera, and are not related to either globe or Jerusalem artichokes.
Ground: The tubers rot in badly drained ground. An easy to work loam is best, in an open sunny position. They are ready for use from late October and can be left in the ground until wanted. So make sure they are planted in a spot which is not water-logged in winter.

Planting: In March or April, 4 ins deep and 9 ins apart. If you grow more than one row, space the rows 15–18 ins apart. They grow to about 18 ins high. If you find difficulty in ordering them from a nurseryman, try a greengrocer who supplies a Chinese restaurant.

ARTICHOKE : Globe

An impressive plant frequently grown in herbaceous borders rather than the vegetable garden, for the beauty of the 3 ft silvery leaves, and stately 4–6 ft high stems crowned with a large bud, producing several smaller offspring on short side branches below. If allowed to bloom rather than being cut for the pot, the buds produce huge purple-blue, thistle-like flowers, much in demand by flower arrangers, either fresh, or dried and petrified in lacquer spray.

Ground: Rich, deep, moist but well-drained, mostly in sun. Dig in plenty of compost, and because this is a permanent crop, staying about five years in the same place, add 1 lb a square yard of bone meal, which is a slow-acting, long-term fertiliser. Heavy wet soil will need lightening with compost or peat, so they are not 'boggy' in winter; light sandy soils will need humus to keep the soil moist in summer when the plants most need it.

Planting: Globe artichokes can be raised from seed if you are willing to take the chance of rearing tough prickly-headed 'rogues'. Sow in an open seed bed in May, prick out to 6 ins apart, and plant out in the autumn. Their buds appear the following year.

Far better to get 'offsets', young rooted plants from nurserymen from March to May. They need armfuls of room 3–4 ft apart, in rows 3 ft apart. Plant them at least 3 ft from the edge of a path or you will be constantly cutting back the overhanging leaves.

Propagation: When a plant has 'settled in' you can multiply your stock by taking suckers from the parent, after it has produced three or four crops.

In April, clear the ground from around the main plant, and with a sharp knife or spade, slice off the outer growing sections with a portion of root attached. Plant them as you would bought 'offsets', 4 ins deep, and water well till they 'take'.

After Care: Spread manure or your chosen fertiliser around the plants every spring – or 2 ozs of dried blood. Like so many silver-leaved plants, globe artichokes are not particularly hardy and need some protection in the worst winter weather. This can be straw or ashes, over the crowns, but remember to take it off before you create a mini-compost heap and rot the plants completely.

ARTICHOKE : Jerusalem
A hardy herbaceous perennial. You eat the roots which resemble potatoes with knobs on. The plants can stay in the same place for about ten years if you want them only occasionally, taking up the delicious roots whenever you need them. After a few years, however, they can become an underground matted entanglement, and the crop smaller, difficult both to fork out and to scrub clean. Consequently, most gardeners replant every year.

Ground: The easiest of all vegetables to grow. It succeeds in almost any soil well-worked but not newly manured and seems immune to disease and pests. Open and sunny position to get largest tubers, though they will grow in any odd corner. In such unfriendly spots, the tubers will be more knobbly and smaller.

Planting: Start with a good modern silver-skinned variety, vastly superior to the inter-twined odd-shaped ones. Choose their site carefully remembering that they will grow to 8 ft tall if you let them and cast shade. Plant in January or February 3–4 ins deep and a foot apart, in rows 3 ft apart.

After Care: If you leave them in the ground for years they spread sideways. They can be controlled by being dug up every year; or used as windbreaks by clipping in June. If not needed for this purpose cut the stems down to a foot from the ground in November, so you can see where they are for digging up and using.

Propagating: Choose the smoothest tubers from the previous crop, about the size of a hen egg.

Forget about raised beds and all advice which makes asparagus sound such hard work and a luxury crop. It is not a demanding plant and with very little attention will produce fat stems for 20 years and more. You have to like it enough though to give up ground all the year in return for a crop lasting about two months.

Ground: Being a permanent crop, the soil must be extra rich, but almost any non-acid type is suitable, particularly sandy ones, which warm up more quickly in spring. The position should be sunny, and the ground well dug and enriched with compost or general fertiliser the autumn before the crowns are planted. All perennial weeds MUST be dug out first, as you cannot get at them once the octopus-roots of the asparagus are in place.

Planting: You can grow from seed or buy one or two-year-old plants. Don't be tempted to order three-year-olds in the hope that you can start cutting them the first year. Their roots are too large and old to stand the transfer; they will either suffer from shock or refuse entirely to take to their new quarters.

They should be out of the ground for the minimum time when being transplanted, with their roots kept moist. If there are delays in rail or post, casualties are inevitable. Plants are sent out from nurseries in April and it is safer to prepare the planting trenches before undoing the package. Then get the job done as fast as you can – with a helper if possible, to hand out the crowns.

You can grow them on the flat, as long as the lower soil is never water-logged. Dig a trench 9 ins deep and wide. Keep a ridge of soil down the middle on which the crown (growing point) can 'ride' with its long roots astride. Then cover the crowns with 3 ins of soil. Allow 18 ins between the plants. Grow in double or single rows at a distance apart to suit yourself. Opinions vary.

Propagating: Seed is certainly the cheapest way of growing a large amount of asparagus. Even if you start off with bought plants, it is a wise plan to begin your own little nursery in an odd corner at once as replacements for failures.

This eventually will give you a longer 'cutting' period, as seedlings vary in behaviour, some coming up earlier or later than pedigree crowns carefully selected by skilled nurserymen.

The seed is slow to germinate, so it helps to soak it for 24 hours before sowing to soften it. Sow in fine soil in drills $1\frac{1}{2}$ ins apart when the soil has warmed up in April and May.

When the wispy little shoots are big enough to handle, thin them to 4 ins apart. They can be transplanted to their permanent homes in the autumn while you can still easily see where they are by their miniature growth. Or mark their position carefully and put them in their final spot in April, just as they are beginning to make new shoots. DO NOT cut them until the third year.

After Care: The usual cutting period is from May up to the second week in June, but if you raise a number of different varieties you can cheat considerably about this nicety. The reason for letting temptingly edible shoots go to seed, is to allow the fern to grow up and send back nourishment to the hard-worked roots.

Be careful to cut down the fern when it has turned golden but before the bright red seeds drop in autumn, or you will have a multitude of seedlings to remove before they choke themselves and their parents.

Weed beds by hand, never use a hoe or you might damage the roots.

When the fern has been cut to the ground in autumn and removed to the compost heap, dress the rows with rotten manure or compost, enough to cover the surface. At the end of February rake off the remains and feed them with a complete fertiliser at 4 ozs per square yard. Work this into the soil lightly no deeper than 2–3 ins so as not to disturb the crowns.

Harvest: Opinions vary about the first cuttings from young plants; some gardeners take two or three sticks from three-year-old plants that have been planted for a year; that is the year after putting in two-year-old plants or the third year after sowing seeds direct in the bed. The longer you can curb your impatience, the more time the roots have to build up

31

their strength and produce really fat stems during the onslaught of anything up to 30 years.

Use a strong, narrow, pointed knife. There is a special asparagus knife, but I use one intended for grapefruit. The trouble is you have to cut the actual shoot which is ready to eat without damaging the adjacent buds well below the surface. Cut about 3 ins below the soil which should be just above the hidden crown. Some people like to cut the sticks as soon as the buds appear, so most of it is blanched, others like to wait until it is green, or purple, depending on the variety.

Whichever way you prefer, remove all the shoots on an established bed, which are ready for cutting, whether thick or thin. If the thin ones are left to go to seed the plant will stop producing fat sticks.

Stop cutting by the end of June, however early or late you were able to start.

ASPARAGUS PEA

A rarity for the gourmet table, seldom possible to buy. Not a true pea, but an annual related to Lotus and known botanically as *Tetragonolobus purpureus*. It grows to 18 ins and needs no staking. The pretty red flowers turn into curious winged pods which must be picked when an inch long, otherwise they become stringy and inedible. A niggling job, but their flavour is unique and well worth the trouble.

Ground: Any reasonable soil which gets plenty of sun. Don't add fertiliser unless the ground is very poor.

Sowing: About the middle of May, in rows 18 ins apart. The seed is very small so should be no deeper than ½ an in.

BEANS, BROAD

The hardiest of the bean family. In some areas and soils they can be sown in November and will stand the winter to provide an extra early crop.

Ground: They are not fussy and will grow on both light and heavy land. It should be well worked and already manured for a previous crop.

Sowing: From late February as soon as the soil is fit to

32

tread on. Make a drill 6 ins wide and 2 ins deep, plant the beans in a double row, staggered so that they are 6 ins apart. Make the rows 2½ ft apart and put a group of extra seeds at the ends as replacements for any gaps.

After Care: As soon as beans start forming from the bottom truss of flowers, break off about 3 ins of the tender top of each plant to prevent blackfly from forming there. They find it difficult to get established on the older, tougher parts of the plant.

If you are too late to stop an attack, apply a derris dust or spray.

BEANS, FRENCH CLIMBING

Also known as Pole beans. It is sown earlier and crops earlier than its relative the runner bean. It combines the distinct and delicate flavour of dwarf French varieties with the heavy cropping of runner beans.

May be grown against a fence, round poles, as a hedge or up trellis or netting against a sunny wall, or even as a hedge.

Ground: Any type. Lightly fork in powdery compost during spring, where they are to grow.

Sowing: From late April to early July. Their enemy is wet ground, not cold. Plant the seeds 2–3 ins deep and 6–9 ins apart in double rows a foot apart. Allow 5 ft between each double row.

After Care: It pays to nip out the growing point when the plants are 6 ins tall, so that they will produce sideshoots, and become more bushy. In any case, stop them growing when they have reached the tops of their supports. Give them frequent waterings during a dry season. A mulch will conserve moisture during warm weather.

BEANS, FRENCH DWARF

Suitable for the smallest gardens, or pots and windowboxes.

Ground Rich, light and sunny.

Sowing: From the end of April in the south, and a fortnight later in the north. Drills should be 2 ft apart and the seeds sown thinly 2 ins deep. They can be in double or single rows, the seeds 6–12 ins apart depending on variety and garden space.

After Care: Water well in dry weather or the beans will be hard and stringy. Take care when picking them, not to loosen the roots. A careless tug at a bean can leave you with the whole plant in your hand.

BEANS, RUNNER

Culture – as for Climbing beans, but sown two weeks later.

BEETROOT

There are several distinct types, globe or round, long, golden and white. The longrooted ones take longer to grow and cook, and are not as attractive to serve.

Ground: Well worked sandy soil with some additional salt, as it is a maritime plant. This is best provided in the form of seaweed or fish manure well before the seeds are sown. They will grow perfectly well in other soils if they have been properly conditioned, are light and well drained, and were manured the previous year.

Sowing: From late April till early June in drills 15 ins apart and 1 in deep. Thin out first to 4 ins and finally to 9 ins apart.

After Care: Never let the roots be short of water or they will grow coarse and 'woody'. They will benefit from a dressing of common salt in July, $\frac{1}{2}$ oz per sq yard, preferably during a showery spell.

BROCCOLI, PURPLE AND WHITE SPROUTING

Ground: This should be prepared the previous autumn with manure forked into the top soil but not buried deeply. They thrive in clay or a heavy loam. Avoid planting on loose, freshly manured soil. They do best on firm ground which was fed for a previous crop. Choose a place where they will not be rocked by winds.

Sowing: From March until May in a well raked down seed-bed, in rows $\frac{3}{4}$ in deep. Thin the seedlings as soon as possible to avoid overcrowding and 'leggy' results. When they are 2 ins high, give them 4 ins apart.

Planting: During June and July transfer them to their already prepared beds, 2–2½ ft apart in each direction, and firmly trodden-in.

BRUSSELS SPROUTS
Culture – The same as broccoli, except that they need more growing time. Sow not later than April in an open ground seed bed. They should be planted out earlier too, in May and June, as the larger the plants are, the longer they take to settle in. Plant with the lower leaves resting on the soil.

CABBAGE
There are distinct types which can be grown to mature at any time of the year.
Ground: As for broccoli.
Sowing: All types in a seed bed. March-April for summer and autumn cropping; July-August for spring crops.
Planting: As soon as there is spare ground. Winter cabbage needs plenty of room, rows and plants 2 ft apart. Compact varieties may be closer. Spring cabbage can follow the space left by early potatoes, peas or beans, 18 ins apart and 12 ins between rows.
Note: The seasons attached to the various types indicate the time of year they are EATEN not SOWN.

CALABRESE
Often listed in catalogues with sprouting broccoli, to which it is related, but has very different habits, and deep green, rather than white heads. It crops in late summer and autumn instead of the following spring, not outstaying its welcome and leaving space for other crops.
Culture – As Broccoli.

CARROT
Though you can have them all through the year, the first ones are the most delicious.
Ground: Friable, well drained and should NOT have been manured for an earlier crop or the roots will split.
Sowing: Soil must be raked to a very fine tilth, like bread-

crumbs. Sow in early April in shallow drills 10 ins apart. On heavy land sow the short-rooted varieties.

After Care: Thin seedlings to 2 ins apart in the rows when large enough to handle. Later remove alternate plants so the roots have 4 ins in which to develop.

CELERIAC

The globular roots have a pronounced tangy taste rather like a celery-flavoured turnip. It can be eaten raw or cooked.

Ground: Extra rich, and preferably prepared in the autumn with manure or compost. If it has to be done close to planting time, whatever is used must be well decayed.

Sowing: In cold areas they have to be started off in boxes early in March and given some protection (cloche, frame, windowsill). Plant out in May or early June 12 ins apart, in flat rows 12 ins apart.

After Care: A greedy eater. Give it weak liquid manure once a week from the end of June. Remove any side shoots which appear.

CELERY, SELF BLANCHING

Culture: The same as Celeriac. Neither needs earthing up, or trenching. Grow plants 8 ins apart in square blocks, covering only the roots.

CELTUCE

A 'two for the price of one' vegetable. It is a form of lettuce, but instead of producing a heart of leaves, it sends up a thick stem 2 ft high, with a top-knot of leaves and a few side ones. These are used as lettuce. The real delicacy is in the heart of the stem which is crisp and a nutty flavour. It can be used like celery, raw or cooked.

Ground: The same as for lettuce, warm open and moist, so that it can grow without a check.

Sowing: In April and May, thinly in rows 12 ins apart and $\frac{1}{2}$ in deep. Thin plants finally to 10 ins apart. At first the seedlings look like cos lettuce, but once they have developed a rosette of fairly large leaves, the stem begins to shoot upwards.

As easy to grow as the dandelion to which it is related, and at a fraction of the cost of the always expensive and frequently tired shop offerings.

Ground: Any, well worked but not freshly manured.

Sowing: Between the middle of May and beginning of June (in their permanent position), in shallow drills 1 ft apart. Then thin the seedlings to 9 ins apart.

After Care: They have the look of lettuce to begin with and during the summer the green leaves can be cooked like spinach. But it is the chicon (the blanched secondary growth sent up by the old roots) which is the delicacy.

In October twist off the leaves, carefully lift the roots and pack them tightly into light soil and keep in the cool until you want to force them. This can be as soon as you like and as many at a time as you want. There are many ways of doing this. Basically it means replanting the old roots close together in damp soil or peat, in boxes, and excluding all light. If you can spare the ground and are daunted by the performance of uprooting and forcing, simply leave the plants where they are (the green leaves will rot in winter) and in spring earth them up with sand, peat, soil or anything you can spare to keep out the light, then wait for their yellow tips to break through. They will not be as perfect, and you will have a lot coming up at the same time – March-April – but it is much less trouble and time-taking.

CHINESE CABBAGE

More like a cos lettuce than a cabbage, but actually belongs to the mustard group of the large brassica family.

Ground: As for lettuce.

Sowing: Seeds ½ in deep in rows 2 ft apart where they are to grow in June. It is disastrous to move them as they immediately try to flower and are useless to eat. You can sow them thinly or in clusters 15 ins apart and later thin the seedlings to leave one in each place.

After Care: They need plenty of water. When the heads are fairly large, tie the leaves together with raffia or soft string to blanch their centres. They can be eaten raw or cooked.

Warnings: They will bolt to seed if sown before June or if the soil is allowed to dry out. They are not hardy and will be killed by frost. They must be used as soon as ready – about 9–10 weeks after sowing.

Advantages: You can fit them between two rows of tall growing vegetables where the soil is shaded and will remain moist.

CUCUMBER, RIDGE

Ground: Sunny, open spot protected from strong winds. Rich, moist and well-drained.

Sowing: Cucumbers resent transplanting and it is best to wait till the soil is warm enough to set the seed where the plants are to fruit 3 ft apart. This can be hastened by warming the ground with cloches, or putting jam jars over the individual spots before and after the seeds are put in 1 in deep, removing them when there is no danger of frost. The seed must not be sown before early May or the plants will be up before the weather is ready for them.

An alternative method is to raise the seeds in the warm in individual peat pots which you plant out early in June. Give them plenty of water and the roots will grow through the pots so there is no disturbance.

After Care: Nip out the growing tips when 6–7 leaves have been formed to encourage side shoots.

To make sure of a heavy crop give a mulch around the roots of well-rooted lawn mowings (NOT any which have been dosed with weed killers!), or old compost or manure. Never let the plants lack moisture or the fruits will stop forming. Be careful not to wash away the soil and expose the roots when watering, or to let water collect round the stems. This could rot them.

GARLIC

As easy to grow as shallots.

Ground: Light, rich, sunny, dry soil which was manured for a previous crop.

Planting: Separate the garlic bulbs into 'cloves' and from February to April plant them 6 ins apart in rows a foot apart, covered by about ½ in of soil.

After Care: They need no further attention except weeding, until their tops turn yellow, usually about-mid-July, at which stage they are ready for lifting. Lay them in a sunny place to dry for a few weeks.

GOOD KING HENRY

Also known as mercury, perennial goosefoot, Lincolnshire spinach and poor man's asparagus. To the botanist it is *Chenopodium bonus-henricus.*

It is a permanent multi-purpose plant, easily grown from seed and unjustifiably neglected.

Ground: Well manured and free of perennial weeds, the plants will be with you for life, if you wish, so choose the spot carefully.

Sowing: In April or May where the plants are to grow, thinning seedlings to 18 ins apart, in rows 2 ft apart.

After Care: They are slow to germinate and when they first appear look frighteningly like docks. But don't do any weeding, wait till they develop their first characteristic leaves which are triangular.

Unlike asparagus, Good King Henry provides three distinct crops if you give it rich soil and time to develop. The young flower buds which form inside the leaves in early spring are boiled for 5 minutes and served with melted butter.

Or the plants can be earthed up in February, and the young shoots cut and cooked in bundles like asparagus. It is best to take only the flower buds the first year, so that the plants can build up their strength for succeeding years. They also produce leaves which can be treated in the same way as spinach but they have a more bitter taste.

Cut the plants down as soon as they start to make seed heads, and give them a sprinkle of sulphate of ammonia. If you keep doing this you should get four or five cuts a year, though not, of course, in the initial year.

At the end of each season cut off the yellowing foliage

and tidy up the bed. Give it fertiliser then or in the spring.

Old plants are easily divided if you want to increase or separate your stock.

HAMBURG PARSLEY

Although the tops can be used as parsley, it is grown for its long, parsnip-shaped roots which may be eaten raw or cooked, and have a parsley-celery flavour.

Ground: It does not need an open site like most vegetables and so is perfect for a small garden, where it can grow close to a row of peas or other taller crop and enjoy the half-shade. As with other roots, it should not be grown on freshly manured ground or the roots will 'fang'.

Sowing: Sow from March to June in well raked soil where they are to grow, as thinly as possible, in $\frac{1}{2}$ in drills spaced 1 ft apart. The seedlings are slow to appear above ground and a few radish seeds sown among them at the same time will soon germinate and show where the rows are so that you can hoe with safety. Thin them first to 2 ins apart and remove the radishes. When their leaves begin to touch, thin them to a final 8 ins. The roots should be ready for use in September.

KALE (BORECOLE), GREEN LEAVED DWARF AND TALL

One of the hardiest of all greens, it is a better flavour after a severe frost. There are many types, mostly with curled leaves.

Ground: It will grow almost anywhere and in any type of soil, but does best in well-dug and rich surroundings.

Sowing: During March and April, transplanting to permanent places from June to September, the dwarf varieties at a distance of $1\frac{1}{2}$ ft each way, the taller ones 2 ft.

KOHL RABI

Similar in flavour to a turnip but with a superior 'nutty' flavour. It will grow in conditions too hot and dry for many vegetables. Although it is used as a root crop, the green or purple 'roots' grow at the base of the brassica stem, and are completely above the soil surface.

Ground: Light and sandy, enriched with decayed manure

40

or anything which will help the soil to retain moisture. Old mushroom bed compost can often be bought locally and this is ideal.

Sowing: April and May in drills 15 ins apart, thinning out to 9 ins.

After Care: None. It matures very quickly so keep an eye on the strange soil-hovering globe roots and use them when they are no larger than a medium-sized orange. Trim off the roots and remove the few sparse leaves but do not peel.

LEEKS

Don't be frightened or put off by the massive pillars you see as prize winners at shows. You can get them that way if you want to make a hobby of feeding and blanching them, but they will be inedible compared with small, naturally grown ones. Easy enough for any beginner.

Ground: Almost any, though preferably soil which is rich and retains moisture but never becomes 'boggy'.

Sowing: Outdoors during March and April, the plants can either be thinned out 9–12 ins apart or transplanted into holes made with a dibber 5 ins deep, 6–9 ins apart. Insert a pinch of bonemeal, drop down a seedling and wash some earth around it. Then leave them alone till they are the size you want.

LETTUCE, LITTLE GEM OR SUGAR COS

There are lettuces for all seasons, conditions, soils and tastes. This type was chosen for its eating qualities (raw or cooked) and its easy growing nature.

Ground: All lettuce do best in moist yet well drained soil which has been manured for a previous crop.

Sowing: Little Gem is half cos and half cabbage in appearance, for use in summer, and must grow quickly to be crisp and sweet. So wait till the soil warms up, then sow thinly and cut off with scissors any overcrowded patches so as not to disturb their roots and leave them shoulder to shoulder apart.

After Care: Water or mulch in really dry weather so that growth is not checked, or they will 'bolt' or grow bitter.

41

There is no difference between either of these, or pumpkins. They all belong to the same family. There are both bush and trailing kinds, some more suitable for winter storing than others. You may find the same variety under 'squash' in one catalogue and 'vegetable marrow' in another. In America they like to use the overall family name of SQUASH.

Ground: Very rich. If there is not enough good manure or compost to prepare the whole bed, take out individual holes 1 ft deep and 2 ft across, and fill these with any moist, rich, humus material.

Sowing: The same as cucumbers. Indoors during March and April or open ground in the last two weeks of May. Put in the seeds on their sides rather than flat, to avoid rotting, and add a few extra ones at the end of the row to replace failures. Allow 3 ft apart for bush varieties and 5 ft for trailers. Pinch out tips of main growths of trailers when they are 3 ft long.

After Care: Keep the plants well watered and the fruit of the 'courgette' type harvested regularly. Even one allowed to grow 'marrow' size will send a message to the plant that it has reached maturity, and it will stop producing new flowers.

ONIONS

Though these can be bought all the year round it is not always possible to get just the size or variety you want at a particular time for a special dish – such as white, cherry-sized ones for a sauce.

Ground: Rich and crumbly for preference. You can use the same spot for endless years, gradually improving it with manure, peat, wood ashes, lime, and so on, so that it does not have to become part of the regular vegetable rotation.

Sowing: In March-April in drills ½ in deep and 8–12 ins apart. Thin seedlings to 6 ins apart for average sized onions and 8 ins for 'monsters'.

Planting: If you have difficulty in raising onions from seed, 'sets' are easier and can be bought by the pound at the right time for planting. They are miniature onions which you plant with a trowel or dibber in March or April the same distance as the final thinning of seedlings. Leave the tips just showing.

After Care: Birds often pull them out of their rows, either from mischief or they may fancy the dry tops as bedding. Push the bulbs back if only a few are disturbed but if it becomes a nuisance guard them with netting or black cotton. Once the green tops appear the birds lose interest. The same trouble happens with shallots.

PARSNIPS

Less effort to grow than carrots.

Ground: Same as for other root crops, well forked over and enriched for a previous crop, but not recently manured or the roots will fork.

Sowing: In March-April in shallow drills $1-1\frac{1}{2}$ ft apart. Thin seedlings 6–9 ins apart. The seed may be slow to germinate, so be patient.

PEAS, GREEN

These are listed in the seasons in which they crop; first early, second early, maincrop, late, round or wrinkled; dwarf or tall. The round-seeded ones are hardier and can be planted earlier but have not the taste of the wrinkled seeds.

Ground: Rich, containing plenty of chalk or lime, deeply dug and manured for a previous crop. The most important factor is that the soil should be able to retain moisture.

Sowing: In flat bottomed drills 3 ins deep and 6–9 ins wide. Space the seeds evenly 2–3 ins apart, then cover to leave the rows flat. The spacing between the rows should be equal to the height of each variety. The very dwarf ones will crop reasonably well without any sticks, but are more prolific (and cleaner) if supported by hazel twigs or one of the many forms of plastic netting.

After Care: Do not thin the seedlings, and be cautious when handweeding not to disturb the tender pea roots by trying to yank out a thistle, chickweed or suchlike growing among

43

them. Snip off their tops and deal with the weed roots when the peas are over.

POTATO, EARLY

The only ones worth growing for a place in the freezer are those which provide the true pomme frittes and those which taste permanently 'new'.

Ground: They will grow in any soil and almost any condition, but the best results are from well dug and fertilised ground, free of lime, as this encourages scabby skins.

Planting: To get the best yield, start off the tubers by allowing them to sprout in a light frost proof room. Egg trays are useful for this. Stand the tubers, eyed end up and, once they have sprouted, plant them in trenches during March and April, 5 ins deep and 12 ins apart, in rows 2½ ft apart.

After Care: When shoots appear, draw a little soil from between the rows and over them. Continue to do this as the stems lengthen to keep the hidden tubers from the light. If they become exposed they go green and are inedible.

SALSIFY AND SCORZONERA

Known as the 'vegetable oyster' because of the delicious flavour of the roots which look like slender parsnips, though with a finer texture.

Ground: Light and not newly manured or the roots will fork.

Sowing: In April 1 in deep in drills 15 ins apart. Thin the seedlings to a final 9 ins apart. The largest roots come from the earliest sowings.

After Care: They are quite hardy and can be left in the ground until wanted, but unlike parsnips, the flavour is not improved by frost. They can be used from the middle of October.

It is best to leave them in the ground till the last possible moment before cooking, because they bleed badly if damaged as they are lifted, and lose flavour.

Differences: Their culture and flavour are similar, though they belong to different families. Salsify has white roots and Scorzonera, black, and for my taste, a much better flavour. It

44

is also said to help in various forms of indigestion. Scorzonera is a perennial and can be left in the ground a second year to grow larger, but this wastes space where it is precious.

Warnings: In a bad winter, even if you are able to lift the roots, you cannot distinguish them from some soils and they get horribly mangled. It is better to lift and use them in autumn before conditions turn nasty.

SEAKALE BEET (SWISS CHARD)

A form of beet grown for its leaves. It is dual-purpose, the thick mid-ribs are cooked like celery, and the glossy leaves as spinach. Ruby Chard is a striking red-stemmed form.

Ground: Any, from light to heavy, preferably rich.

Sowing: As for beetroot, but leaving the plants at least 1 ft apart.

After Care: Keep the plants productive by pulling the leaves regularly. Twist them away from the base rather than cut. Like spinach, it objects to losing too many leaves at one time.

SHALLOT

Grow as for onion sets but the soil need not be so rich. Press them very firmly into the soil 6 ins apart with the tips just showing.

SPINACH BEET

A more reliable crop than summer spinach as it does not bolt. Like seakale beet but without the broad mid-rib.

Ground: As for beetroot.

Sowing: April and again in August in shallow drills 15 ins apart. Thin the plants to 1 ft apart.

After Care: As for seakale beet, taking one or two of the largest from each plant. Over picking and under picking both slow down the production of leaves.

SPINACH

Ground: The annual summer spinach tends to go to seed rapidly on very light, dry soils which must be enriched with moisture-holding compost.

Sowing: From March till mid-July in drills 1 in deep and

45

1 ft apart. When plants touch in the rows, snip out alternate ones for use, and leave the rest to grow on. From these gather leaves as wanted.

SWEET CORN

Ground: Well manured, sunny and sheltered from strong winds which would rock the tall stems.

Sowing: Outside under cloches in the middle of April, or without protection in May. Transplanting checks the growth. Sow the seed in a block rather than a single row so that the pollen from the central flower falls on to the lower and surrounding silky tassels, rather than being blown on to an adjoining row of unappreciative beans or lettuce. Seeds should be 1½ ins deep 2 ft apart each way. Early Xtra Sweet must be grown apart from other varieties. It can become bitter if cross pollinated.

TOMATO : OUTDOOR VARIETIES

Seed is so cheap compared with the bought fruit that it is worth taking a chance on a good summer with bush and dwarf varieties which need very little attention.

Ground: Heavy or light as long as it is well drained and fed. Sunny.

Sowing: Thinly in seed boxes or pots in March in any indoor spot with a temperature of around 60°F. Gradually harden off by putting them outside on 'friendly' days and plant them out towards the end of May, depending on local conditions.

Bush varieties spread sideways and need 3–4 ft all round. Dwarf kinds need 18 ins–2 ft.

After Care: Spread straw under bush types. Take out the growing tips if they are getting out of hand.

TURNIP

Ground: Rich and moist so they will grow quickly and be sweet and tender rather than woody and coarse.

Sowing: From April to July in drills ¾ in deep, 15 ins between the rows. Thin out from 6–9 ins apart according to the size you like.

HERBS, THEIR CULTIVATION

They mostly thrive best on light, fertile, slightly alkaline, well-drained soil, preferably in full sun as most herbs originate in Mediterranean countries. But they will do well in any ordinary, even poor, garden soil as long as it is not acid.

BASIL : ANNUAL

There are two forms, sweet basil and bush basil. Sweet basil has the larger more aromatic leaves, but is also the more tender. It must be started off under glass or a plastic mini-cloche in February or March; put out the seedlings in May, 8 ins apart. Or, if you do not want a large quantity you can keep them in the house permanently in pots. Never allow the small white flowers to form at the ends of the shoots. Pinch them out at bud stage or growth will be slowed down. It can be sown in the open in mid-May.

The dwarf bush basil has the characteristic flavour and scent but the leaves are much smaller, and less succulent. It can be sown in the open in April.

Uses: Transforms all tomato dishes, shredded raw on to salads or any cooked dish containing tomatoes. In sauces for fish and chicken; soups, omelettes, soufflés and herb butters.

CHERVIL : ANNUAL

Unlike most herbs it likes moist and semi-shady conditions. Sow from March to August. Thin seedlings to 12 ins apart. Pinch out flower buds as soon as they appear.

Uses: The young leaves resemble parsley but have an aniseed flavour. For salads, salad dressing, soups, fish sauces, egg dishes.

CHIVES : PERENNIAL

A member of the onion family grown for the delicate flavour of it slender tubular leaves. Easily grown from seed sown in spring, but takes some time to reach usable size. It is quicker and easier to buy some clumps and divide them into smaller ones between September and April. They soon multiply below ground, and every three or four years should be divided and

47

replanted in clumps of about 6 bulblets, 9 ins apart. Clumps should be cut in succession, about 1 in above soil level, to keep up a constant supply of fresh leaves. Never let the purple flowers develop, or you will have woody seed stems rather than leaves.

Uses: Chopped and sprinkled into salads, egg and cheese dishes, soups, creamed and jacket potatoes.

DILL : ANNUAL

Resembles fennel and grows up to 3 ft with delicate feathery leaves. Sow in March or April thinning seedlings 9–12 ins apart.

Uses: A sweet, sharp flavour, excellent for fish sauces, soups, cooking with cabbage and sprinkled on new potatoes.

FENNEL : PERENNIAL

Sow the seeds in March and thin the seedlings to 24 ins apart. The plants will grow 5–6 ft tall if left alone. By keeping the main stem and side shoots cut, or picked back, there will be a plentiful supply of fresh leaves.

Uses: For stuffing fish or making sauces to accompany it. The slight aniseed flavour goes well with pork and veal. Used sparingly the finely chopped fresh leaves may be added to soups and vinaigrette sauces.

MINT : PERENNIAL

There are dozens of different types from common, to spearmint and apple. The best ones for sauce and cooking with potatoes. Plant in moist soil, in sun or partial shade, between October and April. Increase by cutting the roots into 6 in lengths and replanting in autumn or spring. All mints 'travel' and can be restrained in a sunken container such as a bottomless bucket, but they are so shallow rooted and easy to dig up if they stray, that it is not essential.

Uses: Sauce, tea, jelly – too well-known to need further comment.

PURSLANE : ANNUAL

A distinctive and unusual flavour. Sow the seed in spring, thin till the plants rub shoulders.

Uses: The young leaves and shoots are used as a salad, and later in soup or as a cooked vegetable.

SORREL : PERENNIAL
Sow the seeds in early April in shallow drills 18 ins apart. Thin seedlings to 3 ins apart when they are small, and eventually to 18 ins between each plant. They transplant happily if you want to move their bed. Never let them make a flowerhead.
Uses: The young leaves raw in salads. Blanched, finely chopped leaves used to flavour soup. Treat like spinach and make into a purée to serve with veal, pork, fish and eggs.

SUMMER SAVORY : ANNUAL
Sow seeds during April in a sunny well-drained spot, and thin seedlings to 6–9 ins apart. It looks rather like rosemary.
Uses: Use both leaves and flowers in salads, soups, fish and egg dishes, At its best cooked with, or added to, any kind of beans.

SWEET MARJORAM : HALF-HARDY ANNUAL
Though often listed as a perennial, it cannot be trusted to behave as such in this country. (The less pungent Pot Marjoram will oblige.) Sow the seeds in slight heat during March, thin out the seedlings, harden off and put the plants out in May. Or they can be sown outdoors in April and thinned to 6 ins apart.
Uses: Sweet, distinctive flavour; used extensively in Italian cooking. For practically any hot savoury dish, stuffings and omelettes. An invaluable prestige-aid for the cook.

TARRAGON : PERENNIAL
It is grown from divisions or cuttings. A small bought plant will in a year be large enough to divide and increase your own supply. Plant in October or March in good dryish soil.
Uses: Sauces or stuffings for chicken and fish. In creamed soups, particularly tomato. In vinegars for salads and mayonnaise.

FRUIT TREES AND BUSHES:
THEIR CULTIVATION

Apple: apricot: cherry: damson: fig: nectarine: peach: pear: plum.

A good free-working, medium loam suits nearly all fruits best, though apples, pears and plums will put up with many types of less favourable soil; cooking apples being more tolerant than dessert kinds. Dessert pears on the whole prefer warm, well-drained, but not dry soils to bring out the full flavour of the fruit.

For all of them except figs, the places where the trees or bushes are to be grown must be deeply dug and prepared well before the plants arrive. The subsoil of the planting hole must be broken up, forked over, and humus-making organic matter worked in.

Planting preparation will depend entirely on individual conditions, but try and choose an open sunny place, with no overhead shade and NOT in a wind-tunnel. Spread out the roots, shake and firm the soil between them. Put in a stake if necessary before you completely cover the base, so you can see what you are doing and avoid spearing the roots. The trees should be planted with the soil at the same level it grew in during its nursery days.

TYPES AND SHAPES

Standard: The branches spring from a central stem about 6 ft high. A half-standard has a stem only about 4 ft, though the top growth is just as big. These are more suitable for the private gardener, because of the difficulty of harvesting, pruning and spraying the larger standard tree.

Bush: The central stem rises only 2–3 ft, then splits up into main branches. They can eventually reach quite a size (for the average suburban garden) and you will need a ladder to deal with the fruit and pruning. A small garden would be better off with a dwarf or semi-dwarf bush.

Pyramid: A dwarf bush-type tree in which the branches, starting about 1 ft from the ground, radiate from it like a Christmas tree.

Cordon: The best for a limited space, as the tree is restricted to a single stem, either upright or at an angle. Less common are double cordons (shaped like the letter U) and treble cordons.

Espalier: They can be trained on walls or fences, or in the open on a series of wires.

POLLINATING

Many varieties are self-fertile (they set fruit with the pollen of their own blossom). Others need to be cross-pollinated by another variety before they will set fruit. Even self-fertile varieties often crop better if cross-pollinated. As for apples, if nearby gardens grow different varieties which bloom at the same time, you need not give up space for a cross-pollinator, if you are short of it.

Make sure when ordering fruit trees that you not only get the varieties of your choice, but if necessary, a suitable pollinator.

SPECIAL NEEDS

Apricots, peaches, nectarines: The difficulty of growing them outside in this country is not lack of sun, but that frost and cold winds may kill the blossom and prevent pollination. For this reason they are best grown as fan-trained trees against a south or south-west facing wall, sheltered from wind. If there is danger of frost while they are in bloom, give night protection with some lightweight, fine-mesh Netlon.

Plums: They do best on a heavy soil but it MUST be well-drained. There are no really dwarfing rootstocks for them and even a bush type is likely to get out of hand. They also suffer a lot of damage from birds (bullfinches get most of the blame), which pick out the buds in winter. The most suitable way to grow them is like peaches, fan-trained against a wall. They do not make good cordons or espaliers.

Figs: The only disadvantage of the home-grown fig is that it tries here to do naturally what it does in its native sub-tropical home – produce two or three crops a year. We have to restrict it to one, by restraining the roots in a strait-jacket, to stop them producing more leaves than fruit.

This can be done in several ways. Make a hole 30 ins wide

and deep and line it with bricks, leaving a few holes in the bottom for drainage, these being well covered with tiles or crocks to prevent the roots from growing through. Alternatives are corrugated iron, or an old water tank with the bottom pierced or removed.

They need a sunny, sheltered spot, backed by a wall. To keep growth under control, tip all young shoots and remove dead wood in April when the buds are just breaking. Later, nip out the growing tip of all mature shoots when they have produced five or six leaves, usually in July or August depending on variety. You can also take liberties with surplus shoots by removing them in summer.

The most important point to understand is which figs are ever likely to ripen, as the tree produces embryos at all stages. Any sizeable fruit remaining on leafless branches in autumn must be picked off. They have developed too late in the summer to ripen and are too large to survive the winter. They will only weaken the tree by drawing sap and may contact mildew if left.

The figs which will ripen the following summer, should, by the previous autumn, be the size of a small pea. Those that begin to get fat too early in the year should also be removed, as they will be frost bitten. These usually develop on sappy, too vigorous growth. Figs all bend their necks, as though for the axe, when they are ripe for picking.

Apart from the careful planting, they need less attention than most fruit – no spraying for pests or disease; not pecked by birds; almost no feeding or pruning. Their only enemy is the wasp – you have to get there first.

SOFT FRUIT: THEIR CULTIVATION

CANES

In general they will never make any headway in heavy clay or chalk, but are otherwise easy to grow if the soil is well-drained, rich and reasonably sunny. Prepare the ground thoroughly beforehand, by forking it over, removing all perennial weeds, and lightly forking into the surface peat, leaf-mould or farmyard manure, to retain moisture for their shallow roots. After the canes are planted, a layer of one of these spread over the surface soil will do a power of good, with a fresh layer applied every spring.

No chemical fertiliser should be given until after the first growing season, but as well balanced product should go on annually, in winter, or early spring. Never give them lime.

CANE TRAINING

All garden brambles, except autumn-fruiting raspberries make new growth one year to fruit the next. Therefore the young canes must be nursed and protected from their elders, whose vigorous foliage, blossom and fruit might keep them in the shade. The young canes will not bear well unless they have been ripened in the sun in their first year.

Another hazard, particularly with rampant growers, is that the new canes get stepped on, or fallen over, when you are picking the fruit.

There are several methods of keeping the two age groups apart. Young shoots are tied closely into the centre, and as they grow long enough are trained along the top supporting wire, with the fruiting canes spread on either side like a fan. Or you can make a centre parting, as it were, with new growth on one side and fruiting canes on the other.

The weaving method is usually reserved for the stronger varieties such as blackberry Himalaya Giant, which can form canes 15 ft long in one season. These you interweave up and down, sideways along the wires.

Always wear gloves and beware of their vicious thorns. You may have to protect your eyes when tying in the whippy snake-like branches.

The thornless varieties are less painful to deal with and consequently get more attention. They are certainly safer near a path, but they have not quite the same flavour or vigour of their fierce relatives.

BLACKBERRY

A single plant is quite enough for most households. They will do you the favour of hiding or disguising a garage, wire boundary fence, nasty wall, etc., as they produce canes up to 12 ft long annually. If you have room for more than one, space them 6–9 ft apart.

Plant in autumn and prune them back at once to a fat bud 12 ins or less from the ground. The canes bear berries in their second year, after which they should be cut out entirely, and the new canes which have been growing up during the summer, are trained in as replacements. Nine or ten canes are enough to leave on one bush to develop. Keep them well fed and watered in summer, and after the initial season, one plant can produce 10–15 lbs of berries from September to October.

BLUEBERRY

The cultivated blueberry is a hybrid variety of the wild bilberry or whortleberry. The berries are much larger, are ready for picking in August-September and have the characteristic blue-grey 'bloom'. The bushes grow 4–7 ft, which makes the berries easier to pick than the ground-hugging wild ones, and they live for 30 years or more.

The problem is that they need conditions which no other fruit would tolerate for a moment – an acid, boggy, peaty soil in partial shade. Just the kind of place where rhododendrons and azaleas are most at home.

They need to be planted deeper than most fruit bushes to encourage sucker-like shoots below soil level and build up a large plant as soon as possible. March is the conventional planting time, but as they lose their leaves in winter, there is no reason why this should not be done from November till early April, if the moist soil you choose for them is possible to dig.

BOYSENBERRY

My favourite cane fruit, the shape of a huge glowing mulberry, with a taste closer to loganberry than blackberry from which they are jointly bred. They give the sensation of biting into something exotic. The very large, almost pipless berries at first turn red but should not be picked until almost black

– August to September, depending on district and conditions.

They resist drought better than most cane fruit. Grow and train them as for blackberries.

CURRANTS, BLACK

Need a deep, sandy loan, well-drained but which holds moisture in spring and summer. Will tolerate worse drainage than most soft fruit but avoid areas likely to be waterlogged in winter.

Planting: Give them plenty of room, 5–6 ft between bushes. This may seem extravagent when the plants are young and small, but you can grow salad crops between them, and in a very few years they will be large bushes. THEY must have light and air and YOU must have space to pick and prune. You will get a larger and better quality crop from two properly planted bushes than from half a dozen crowded and neglected.

Plant firmly, between October and March, and slightly deeper than the soil level in the nursery, so that all the shoots arise from the soil.

Pruning: Black currants produce most of their fruit on one year old wood. Cut them down to 8 ins above the ground as soon as they are planted. No fruit will appear until the second summer, and when this has been picked, cut out the stems which have produced it – leaving the fresh new shoots to fruit the following year, plus all those growing direct from the roots, which will help to strengthen the bush and supply it with new growth.

CURRANTS, RED AND WHITE

These are colour and flavour variations of the black currant and need the same growing conditions, but different pruning because they grow differently. White currants are the sweetest of all. Perfect as a dessert.

Red and white currants produce their fruit on spurs of old wood. In later summer shorten all side growths to three or four buds and cut leading shoots to about half their length. They can be grown on a short 'leg' or trained as cordons.

55

GOOSEBERRY

Grow and give the same pruning as for red and white currants.

JAPANESE WINEBERRY: (OR CHINESE BLACKBERRY)

Grow as for other hybrid berries (blackberry, boysenberry, loganberry, etc.) It is one of the most handsome wall plants; also for covering trellis or wires. The canes, when established, grow to 10 ft or more in a season, and remain a rich crimson through the winter, covered with soft red bristles. The leaves are large, pale green on top and downy beneath. The most important part, the berries, are of medium size and round; an unusual golden yellow turning to wine-red when fully ripe – July to August – with a sweet, unusual flavour.

LOGANBERRY

Some doubt about its parantage which is supposed to be a mixture of an American blackberry and a red raspberry. Grow and treat as for blackberry.

RASPBERRY

Cut newly planted canes to 3–4 buds from ground level, to give the roots a chance to get fighting fit for the many seasons to come. In following years, after the fruit is picked, you cut out the old canes and tie in about six of the strongest new ones.

Note: Autumn fruiting and some new varieties need cutting down at different times depending on what you want them to do, and when.

RHUBARB

Strictly speaking, this comes under vegetable growing but as it is invariably eaten as a sweet rather than a savoury I include it here.

It survives in all sorts of unexpected places, but for a first-class crop you want an open site and well fed soil. It can be grown from seed if you are patient and want fields of it. Otherwise buy young crowns in spring or autumn and plant 3 ft apart.

Blanching: For the first delectable pink tender sticks you have just to cover the roots in early January with anything you can get hold of cheaply, which will not blow away. First put bracken, straw, or any packing material round the crowns and pen them in with upside down wooden boxes, old buckets, drain pipes, etc. Alternatively, the packing material can go on the outside of the forcing pots or boxes to keep them warm.

STRAWBERRIES

If your garden happens to be in a frost pocket, avoid early varieties as the blossom may be damaged by frost. They grow in most soils in sun or partial shade but are most prolific if the ground is enriched with organic matter which will retain moisture in summer.

Before planting, fork over the top 6 ins of soil, remove weeds and add well-rotted farmyard manure, compost or coarse peat. No artificial fertiliser is necessary. Firm the ground and leave to settle for at least a month. The texture should not be too fine.

Planting: This can be done in spring, late summer or autumn, set 12–18 ins apart in rows not less than 2 ft apart. Plant firmly, with crown (or neck) of the plant just on ground level. Those set in autumn may be lifted out of the ground by frost. If so, replant them in early March.

After Care: It is advisable to remove the flowers which form in May on young plants so that they can build up their strength for a heavier crop the following year. With summer fruiting varieties, IMMEDIATELY the crop is picked, cut off all the leaves from each plant (with shears or knife) and burn them. This destroys any disease or pests on the old growth and the new will quickly appear.

Runners: Cut them off close to the parent plant as soon as they appear, unless you want to keep one or two from each for new plants. They must, though, be vigorous and absolutely free from disease or virus. Peg them down in June-July, and when they are rooted, transplant to a freshly prepared site. But because of the readiness of strawberries to take to their sick beds with virus disease, it is safer to look on them as a short term crop, to be renewed at least every three years and preferably two.

Perpetual fruiting varieties: These do not usually produce many runners, but those that do bear both flowers and fruit. They should be left on to increase the total crop. They bear fruit from June to October and often later. Their leaves should be removed annually when they have finished fruiting in autumn.

Alpines: Produce no runners and are grown easily from seed. The small fruits are delicious though a fiddle to pick and are produced freely from early summer. They will grow in partial shade and also seed themselves. The plants are best replaced with new seedlings every two or three years. They need rich soil.

YOUNGBERRY

Another cross between the loganberry and blackberry made by a Mr Young, and known in America as the Young Newberry. Fruits are soft, rounding and 1 in in length. Very like a Boysenberry, they are purplish black when they mature in July-August. Canes are moderately vigorous but will yield 10–15 lbs of fruit a plant. Grow and train as for blackberry.

Part Two

HARVESTING

More produce is ruined by wrong harvesting and by the cook, than ever in the growing. It is almost impossible to pick or dig up vegetables when they are too young, though they are so often left till they are past their prime. The reasons are usually a combination of time and greed. It takes more effort to harvest small young vegetables, and you need more of them. They also take longer to scrub, pod or peel. But as for flavour, I have never heard anyone admit they prefer old broad beans with a black 'eye', tasting like leather buttons filled with starch, to the delicious young ones which only half fill the pod.

However, I do know some good-food eaters who prefer a blown out marrow to courgettes. But the former have such a high water content they would hardly be worth space in the freezer.

Some produce is wasted, too, because it is not realised that a number of vegetables are dual-purpose and a few treble, so that almost all the plant can be eaten at different times.

Fruit, on the other hand, has a peak flavour when just ripe but not over-ripe, varying with different varieties, and the use to which you put them.

A freezer cuts down waste. Peas, beans, sweet corn have only a short period when they are sweet and succulent, and if left in the garden will turn tough, stringy and starchy in a matter of days. Anything not needed for immediate eating can be rescued for the deep freeze. The same applies to quick growing summer vegetables. Lettuce can be puréed for soup, and summer spinach 'caught' before going to seed. All the not quite up-to-the-mark can be puréed for use on their own or in soups and stews.

Never pick more than you can prepare and get into the deep freeze within a few hours. This can mean one long session, or a few short ones every few days. This is up to you and your own convenience. If you do pick more than you can deal with in a few hours, or get diverted, some crops will forgive you and 'rest' in the cool of the refrigerator till next day, but peas, beans, asparagus and sweet corn are better under-cooked, strained in cold water, and served with melted butter or a sauce during the next few days.

Always choose a dry day for picking soft fruit to keep its flavour.

Freeze only what you have need and space for. Both good and poor quality foods come out of the freezer no better nor worse than they went in.

VEGETABLES

ARTICHOKE, *Chinese*
The little twisted tubers can be dug as required from November to February. They must be used at once as they discolour if exposed to light and air. Wash them immediately as their curious shape makes it difficult to clean them if soil is allowed to dry on their skins. Never attempt to peel, it would end in tears. They are like little fingers turned into improvised cork screws.
Globe: Large central heads must be removed before the base swells and resembles a cottage leaf. Small side buds are cut when still tight-closed, to be used whole.
Jerusalem: As for Chinese, or peeled and sliced.

ASPARAGUS PEA
Pick when no longer than 1 in.

ASPARAGUS
It is a matter of tast whether you like blanched or green

shoots. For blanched ones, wait for the tips to grow 2 ins above the soil, then carefully feel down their sides with a sharp knife and cut them off about 3 ins below the surface. Care is necessary because new buds will be coming up all the time from the hidden crown, and you may slice them off. Green sticks can be cut at soil level.

Cut everything of the right height whether thin or thick, wash and grade them into sizes.

BEANS, *Broad*

A treble purpose plant. Break off three or more inches of the growing tips when the bottom truss of flowers begins to turn into embryo pods. Use them as spinach.

Baby pods can be picked before the beans are properly formed, as you would mange tout, to use whole.

Always pick broad beans young before they go starchy. If you feel it an extravagance to throw away the tender velvet-lined pods, slice some of them to cook with the beans and add a taste and texture contrast. But the pod must never have reached the stage when there is an encircling string.

French Climbing and Dwarf: When young enough to be stringless.

Runner: A matter of taste whether used as French beans or allowed to grow older, then sliced.

BEETROOT

A dual-purpose plant. Take a few leaves from each plant at a time to cook as spinach. The leaves of Golden beet have the best flavour, superior to summer spinach.

Pull the roots when about the size of a golf ball. The white variety has the sweetest flavour.

BROCCOLI, SPROUTING

Wait until the shoots are about 9–12 ins long, with the buds showing colour, but before they start to open. Then snap off about a third of each shoot. More and more will form later but become progressively thinner and shorter until they are too wispy to bother with. Always break off only a part of each shoot.

BRUSSELS SPROUTS

Remember to gather from the base of the stem, taking a few of the largest from each plant. Press downwards to snap them off. Any leaves in the way can be removed at the same time.

The leafy tops can be cut off and used as cabbage when the best of the button sprouts have been taken, and the 'blousy' sprouts allowed to shoot. These, broken off while still in tight buds are a delicious novelty, (never on sale because only a home gardener would bother to pick them).

All that should be left of a sprout plant is a naked stem, which, chopped into pieces, goes in its turn to feed the compost heap.

CABBAGE

When they have a firm heart – according to type.

CALABRESE

Towards the end of summer they produce a large central head like a compact green cauliflower, which you cut whole. This is followed by a number of side shoots, to be picked with about 6 ins of stem that sometimes need skinning.

CARROT

Pull the largest first, as young as you can bear to take them. Firm the ground around those you disturb.

CELERIAC

Lift the swollen roots carefully from November.

CELERY, SELF BLANCHING

Start using the plants from September. As each is removed fill the space with straw or some material so that light cannot ruin the blanch of adjoining plants. It is not hardy and must be used before hard frosts arrive.

CELTUCE

The top-knot of leaves can be used as lettuce. The real

delicacy is the 2 ft stem. Remove any leaves, peel and cut into long strips.

CHICORY

Whether blanched during the winter or left in the ground and used in spring, cut the chicons as low down as possible, just above the root and just before they are to be used. Exposure to light ruins the flavour. During summer the green leaves can be cooked like spinach.

CHINESE CABBAGE

When it has 'hearted-up', cut whole and use as lettuce or cabbage.

CUCUMBER, RIDGE

Pick them before they begin to swell, when they take more interest in seeding than producing more fruit.

GARLIC

Lift them when their meagre onion-like tops turn yellow. Shake off soil, expose to sun for two weeks to ripen, but keep dry. This can be done by laying them on sacking and moving them into shelter at night, or bringing them out only when it is sunny. Another method is to tie them in small bundles and hang in a sunny shed or on a window ledge, turning them occasionally.

GOOD KING HENRY

If the plants are earthed-up in February, the new shoots are blanched and can be cut like asparagus as they begin to poke through the surface.

If not earthed-up, the young flower buds which form inside the leaves in early spring can be picked, boiled for five minutes and served with melted butter.

The leaves may later be treated as spinach, but are slightly too bitter for some tastes.

HAMBURG PARSLEY

The tops can be used sparingly as parsley. The roots should

be ready for use in September, and should be taken up as and when you want them till they start into growth again the following spring.

KALE
The tops can be cut off early in the year to encourage more shoots. Don't take the complete head as the outer leaves will be bitter, just a cluster of small young leaves. From February onwards you should be able to harvest the shoots as you would broccoli, snapping off a third of their length, to allow more to form below.

KOHL RABI
Must be used when no larger than 2 ins in diameter. The stems will swell to a much larger size, but become inedible. When very young they may be scrubbed and cooked whole. Middle aged and ancient specimens are better peeled thickly and treated like turnips.

LEEKS
Begin using them as soon as they are large enough to serve your purpose, and continue to do so until they have reached maximum size.

LETTUCE
Cut them with tight, solid, quickly-grown heads. Those which have struck a drought or cold spell will be bitter and more appreciated by the compost heap than the eaters.

MARROW AND SQUASH : COURGETTE ETC
Cut off when a few inches wide or long. Never let them grow old enough to get thick-skinned.

ONIONS
The first sign of maturity is when the foliage starts to topple over. This will depend on variety and weather conditions. Help them to ripen by tucking the collapsing leaves neatly between the bulbs. When the roots die, the bulbs can be lifted, and used straight away, or dried in an airy place.

PARSNIPS

Lift roots from October but better flavour is achieved if left till the frost has got at them.

PEAS, GREEN

Pick them while there is still some 'give' when you press the pod, and the peas are not large enough to have taken up all the inner space.

For perfection, pick no more than you can shell and use within two hours.

PEAS, MANGE TOUT

Remove the pods before any peas have formed. The pod should always be flat. If the peas have begun to swell, the pod will have started to develop 'strings' round the outside, and be most uncomfortable to eat.

PEAS, PETIT POIS

These peas are the true, small French variety with a delicious flavour. Must be picked young. Shell them if you have time and want a sedentary task. I boil them in their jackets, then shake out the peas.

POTATO

Early varieties can be lifted as soon as the tubers are large enough to cook and eat.

SALSIFY AND SCORZONERA

Start to use the roots from the middle of October. There are two reasons for this: the roots bleed badly if they are broken or damaged as they are lifted, and they lose flavour.

SEAKALE BEAT : (SWISS CHARD)

Twist a few of the outer leaves from the base of each plant regularly, to keep them productive.

SHALLOT

As for garlic.

As for Seakale beet.

SPINACH : SUMMER

Use thinnings whole, then gather as for seakale beet, after pinching out central growing stem to delay the moment at which they will go to seed.

SWEET CORN : USUAL TYPES

When the silks at the top of the cobs begin to dry and turn black, ripening has started. To test when ready, part the sheath with thumb nails at its fattest part. The grains should be pale yellow but fully developed. They turn golden during cooking. Another way is to 'pop' a grain with your thumb nail. The liquid inside should be like clotted cream; neither runny nor cheesy. As this operation spoils the appearance of the cooked cob, it is better done at the top than in the fat middle, but remember that the grains at the base will be more mature. With experience you can tell by the state of the silks and the feel of the cobs when they are perfect for eating, without having to expose any grains.

Break them off with a downward pull and cook them within an hour at most.

Unusual type: Early Xtra Sweet, an American introduction. Instead of the flavour declining from the moment it is picked, like other sweet corn, it improves and after four hours it is four times as sweet as normal varieties.

TOMATO

A matter of preference, whether you like them semi or fully ripe and the use for which you want them.

By the end of September they are unlikely to go on ripening out of doors. Pull them up and hang them in a dry, frost-free place to continue ripening, or you can cut them off or pick them individually and ripen in the house. Another way is to lay them on newspapers, on apple storing trays, or keep them in dark drawers. By using different methods and temperatures you can have them still turning from green to red at Christmas. They will ripen without sun or heat as long as they do not get frosted, but will ripen faster if they get them.

TURNIP
Start pulling as soon as they are large enough to eat, after about 8–10 weeks. Never wait until tennis ball size or they will have developed coarse fibres and coarse flavour.

HERBS : SEE ALSO CULTIVATION

General advice. Never let them start to make a flower head or they will go to seed and stop producing the aromatic leaves. If you have too many maturing at the same time to deal with by hand, take the shears to them and cut off their heads. They will soon thank you with fresh young growth. Generally, the more mature leaves have the fuller flavour.

Choose a dry, still day and gather the herbs after the early morning dew has dried, but before the sun becomes hot.

Handle the leafy shoots carefully so as not to bruise them and lose some of the aroma.

FRUIT : SEE ALSO CULTIVATION

APPLES
They are ripe when the stem parts from the branch, given a gentle upward tilt.

APRICOTS, NECTARINE, PEACH, DAMSON, PLUM
When the fruit comes away easily from the tree leaving the stem behind.

BLACKBERRY, BOYSENBERRY, LOGANBERRY, YOUNGBERRY
When the fruit breaks easily from its cradle, but before it is in danger of falling off or rotting. Apart from loganberry which stays red when mature, the other berries are not ripe till they turn a purplish black.

BLUEBERRY
The berries are about the size of a blackcurrant, and ready for picking in August and September. They are blue-black with a grey-blue 'bloom'.

CHERRY
Depending on variety, preference and use.

CURRANTS
Always a certain amount of wastage as those nearest the base of the truss ripen first and may begin to fall off before those at the tip are ready to pick. Take off the little bunches whole and shred off gently with a wide-tined table fork.

FIG
When their necks sag and they are no longer upright.

GOOSEBERRY
Depending on variety, preference and use.

JAPANESE WINEBERRY
The round berries turn from golden yellow to wine-red when fully ripe, about July-August.

PEARS
Depending on variety, preference and use. They can be ripened on the tree. Picked and stored till ripe, or used unripe.

RASPBERRY
Unlike the other berries, they must be gently pulled from their central 'plug' rather than broken off. If you have to squeeze hard to achieve this, they are not ripe, wait until they come away willingly, but do not actually fall into your hand or they will be over-ripe and 'winey'.

RHUBARB : SEE CULTIVATION
Hold the sticks close to the crowns and give a sharp pull.

They will come away with a white spoon-shaped base which is edible but needs trimming. Never break or cut them.

STRAWBERRIES
Depending on preference and use. Alpines should be fully ripe for best flavour.

TREATMENT OF RANDOM HARVEST

BILBERRIES
Found on moorland and sheltered places on mountains from July onwards. Freeze as for Blueberries.

BLACKBERRIES
Hedges, scrubland, etc. Freeze as for cultivated varieties but keep a look out for maggots. They freeze well but neither improve the taste or look of the fruit.

COCKLES
They are found close to the surface of the sand on tidal shores about 2 ins down. They are easy to gather by raking or with the fingers if the sand is not too gravelly. As the cockles draw in sand with the plankton they feed on they need careful cleaning. Wash off sand in several waters, then leave them in salted water for up to 12 hours to give them time to expel the sand.
To Freeze: Put them into a saucepan with just enough water to cover the bottom. Cover with lid or towel. Put on high heat and shake pan constantly to prevent them from burning. As soon as shells open they are cooked. Strain liquid and freeze separately from the fish, as when you use them the cockles will need no further cooking, while the liquid will be used in various sauces for them. Home cooked cockles bear no resemblance to the bottled or fish stall ones which lose their juices and plumpness in the mass cleaning and cooking process – they are mechanically scalded open, tossed out, then preserved in brine.

ELDERFLOWER

The large creamy flat headed flowers are abundant in hedges and woods in June, and can be made into 'tea'. They add a musky flavour to water ice, ice-cream, or jellies. The aroma is so strong that a bunch of flowers drawn through any fine jam just before bottling, will scent it deliciously.

To Freeze:

1. Put dry into polythene bags. Use while still frozen or the blooms will go black, though still retaining their scent.

2. Cover with cold syrup. (*see recipes*)

ELDERBERRY

Black juicy fruit in early autumn, insipid raw, but makes distinctive flavoured wine and can also be used in pies and jellies. The fruit is ripe when the heads hang down their necks.

MUSHROOMS AND EDIBLE FUNGI

The common field mushroom is dainty pink and white when young, the gills turning brown then almost black with age. They are found in pastures where cattle have grazed and have a more delicate flavour than cultivated mushrooms..

There are, however, a vast number of other edible fungi to be found in different places at different times of the year. 'Horse' mushroom, the common 'Puff Ball', the 'Fairy Ring', Bluet or Bluetail, the Ink Cap or Ink Horn – to name just a few. But never eat any unless you are certain they are safe. There are only a few poisonous ones in this very large family, but it would be foolhardy to take a chance. There are many illustrated books on edible fungi to help you identify them. Only freeze young, fresh ones. Pack them dry and raw and cut off the stems if they are an awkward shape. They can be used separately in stews or to flavour stock.

MUSSELS

Found where there are rocks for them to attach themselves. Those which live high up on the shore remain small as they can feed only when covered in water. No mussel is poisonous in itself, but those clinging to piers and sewage pipes are not edible because the water is impure, not the mussel. Keep brushing them till the water is clear. Once they are clean, never leave them immersed. Put them in the cool with a damp cloth over them and cook and freeze as soon as possible. *Method:* As for Cockles. If used hot, they must never be allowed to boil or they will go leathery. They can be added at the last moment to a cooked hot sauce.

SAMPHIRE

A little known marsh plant which was served in the past with marsh mutton. It grows in bare mud or rough shingle in salt marshes which are periodically submerged by the sea all round the coast. Samphire is a succulent with jointed much-branched stems and looks rather like a miniature Christmas tree. It needs very thorough washing. Boil or steam the young shoots and serve like asparagus with melted butter.
To Freeze: (a) Blanch for 2 minutes and freeze whole.
   (b) Cook until the outer succulent part can be stripped from the thread like 'ribs'. Discard these.
To use: The pulp will need no further cooking, and can be eaten with cold fish, brain or egg dishes, or with the same food hot and melted butter poured over the samphire.

DEEP FREEZING (AND COOKING)

CAUTIONARY NOTE

Although I shall give the recommended and tested blanching times of the 'popular' vegetables, I have found that they are far too long for my own fresh-picked and young crops, which become completely cooked in the 'approved' time. Even a day makes a difference of minutes to the time it takes to blanch a cauliflower.

The published charts seem to be based on shop-bought produce which may be several days old when purchased. Try a small, experimental spoonful before you ruin a whole batch. If they are cooked rather than 'blanched' in the given time, label them so that you never re-boil them when out of the freezer, but let them de-frost, then toss them quickly in hot butter to get rid of the moisture. Alternatively, put them into a cooked dish at the last moment, giving them just time to heat.

Vegetables are as simple to prepare for the freezer as for the table. They can be cut, cubed, sliced, shredded or left whole. The choice is yours. Root vegetables can be frozen whole when small or cubed. They all need to be washed, then blanched quickly in small quantities in rapidly boiling water. The time varies from 1–2 minutes for such things as baby French beans, to 5 minutes for some root vegetables. After blanching they go immediately into icy cold water, then drained and packed into containers. These are the basic rules. Now to individual treatment.

BLANCHING

Vegetables have to be blanched (scalded) to destroy the enzymes which would otherwise cause deterioration in flavour, colour and texture during freezer storage. It also helps to retain Vitamin C. The same water may be used up to 7–8 times, to build up its vitamin content. The water can be topped up from a boiling kettle. Then start again with fresh water.

For best results use 6 pints boiling water to each 1 lb of produce and never attempt to blanch more than this amount at a time. Count blanching time from the moment the water returns to the boil, then cool immediately under running cold water, or ice-cold water (or a succession of each) to prevent over-cooking and cool as quickly as possible. They usually take the same length of time to cool as to blanch.

Note: If you are caught hopping with more vegetables than you can blanch, you can pop them into bags, just as they are, and deep freeze for up to two months. But this is not a good idea except in an emergency. You will probably forget them and they deteriorate rapidly.

Blanching equipment: There are special Blancher Baskets which can be plunged with their contents straight into boiling water, then still in the same container dunked in icy water till cold. You can also use chip baskets or loose linked lettuce shaking baskets.

74

Having tried all of these I find that even with shaking the baskets constantly, first in boiling then in icy water, the vegetables neither boiled nor cooled fast enough because of the limited space in the basket.

My personal method is to have two boiling rings on high with large pans of water; two spare empty pans at the ready and all the colanders and strainers I have, near at hand. With a double sink you can keep one full of icy water and leave the other to quick-cool under a running tap, and take off the instant heat, before finally cooling.

With this method you do not need pans holding more than two pints of water, and it also eliminates the risk of tennis elbow.

When the water boils, put it in a few handfuls of vegetables, replace lid to contain heat and bring back water to boiling as fast as possible. Start counting, and when time is up pour contents into strainer over empty pan of similar size. Put strainer (colander) under cold tap while you pour blanching water back into hot pan, bring to boil and start with next batch. Put strainer into sink of icy water and remove it before the next batch arrives.

I use two burners at a time because it is a tedious business waiting for just one pan to go through the process. Unless you can start your batches off at regular intervals, you will need two separate timers with distinctive sounds, or a very good memory and a watch with a second hand.

Steam blanching: Put enough water in saucepan to prevent it boiling dry. When water is boiling fast, put vegetables into wire basket or cheesecloth bag and lower into steamer. Cover with lid.

Count blanching time from moment steam begins to stream from the lid.

Steam blanching usually takes half as long again as blanching in boiling water. It is not advisable for leafy green vegetables such as spinach, as they tend to mat together and will not be evenly blanched.

Unblanched for short term storage: Some vegetables, such as peas, and runner and French beans can be prepared as for cooking – podded, topped and tailed or sliced – and frozen

75

without the precaution of blanching, for use within three months. Experiments are still going on with this method and it is not officially recommended. Some housewives claim that the unblanched ones have a better flavour. Some experts say there is a loss of texture, the vegetables can lose colour, and have not yet committed themselves about safety and food value.

If you want to experiment try both methods and form your own opinion. A great deal depends on the speed at which the vegetables can be frozen.

ARTICHOKE, CHINESE

Lift from late October to February. Only take up what you can use at once as the tubers discolour if exposed to light for too long before cooking. Wash immediately as their curious shape makes them difficult to clean if the soil dries on them.

Leave whole – they are impossible to peel. Blanch one minute. To cook, they can be parboiled, then sautéed, steamed then fried in butter till slightly brown, or used in many other ways. (*See recipes.*) A common error is to overboil them. They should always remain firm and almost crisp.

ARTICHOKE, GLOBE, WHOLE

Remove outer and lower coarse leaves, trim tops nad stems. Wash thoroughly under running tap (earwigs lurk between the leaf-scales), or leave in salted water for up to an hour.

Add a little lemon juice to the blanching water to keep their colour. Boil a few at a time 7–10 minutes, cook, and pack in rigid containers.

Globe hearts: Clean and boil whole artichokes till the leaves come away easily. The time will vary with age and size, but it usually takes between 20–50 minutes. Drain, pull off all leaves, scoop out choke with a spoon and discard. Cool hearts and pack, separated by waxed paper, in rigid container.

ARTICHOKE, JERUSALEM

Peel and cube or slice, put at once into cold water with a dash of lemon juice to keep them white, and leave them there

76

till you are ready to blanch, for 1–2 minutes, then cool, dry and pack.

Alternatively, they can be cooked whole for about 15 minutes until soft, the skins pulled off (much easier than peeling them raw) and the insides liquidised into a purée.

ASPARAGUS

Wash and grade the stems according to thickness. Break off the woody base. Don't tie in bundles. Blanch thick stems 4 minutes; thin ones, 2. The very thin 'sprew' stems freeze particularly well as they have little water content. Break them into short pieces and blanch for 1 minute.

ASPARAGUS PEA

Blanch whole for 1 minute.

BEANS, BROAD

Leaf tips. Blanch for 2 minutes, cool and squeeze out excess water with hands as you would spinach.

Baby pods. Blanch whole for 3 minutes.

Shell and blanch young beans for 3 minutes.

Old beans which have escaped you and grown leathery jackets can be shelled, blanched for 4 minutes, the outer skins slipped off while they are still warm, and the tender inner beans put straight into polythene bags and frozen, without the icy water treatment.

BEANS, FRENCH CLIMBING, DWARF, AND RUNNER

Use only small ones. Top, tail and string if necessary. Blanch 2–3 minutes.

BEETROOT

Choose those under 1 in in diameter for freezing whole. Larger ones should be sliced or cubed. Blanch 5–10 minutes, rub or peel off skin, cool in water and pack dry.

Large beet: Cook till tender, remove skin and cut as wished. Beetroot may become rubbery with short blanching and long storage.

BROCCOLI, SPROUTING
Trim off woody parts and large leaves. Cut into small sprigs and grade the sizes. Blanch 3 minutes for medium stems, 4 minutes for thick ones.

BRUSSELS SPROUTS
Use small, tight heads, trim and remove any damaged or yellowing leaves. Wash thoroughly. Blanch 3–4 minutes, cool and drain completely.

CABBAGE
Use only young and crisp specimens. Wash well, shred fine, and blanch for 1 minute.

CALABRESE
As for broccoli, but the stem may sometimes need peeling after blanching.

CARROT
Blanch small youngsters whole, for 4 minutes; cool, then rub off skins if necessary.

Old carrots. Peel, grate, slice, or dice, and blanch for 3 minutes.

CELERIAC
Peel, grate, and blanch for 1 minute, or dice or slice and blanch for 2 minutes. Can be used in salads and stews. Alternatively, cook till almost tender, peel and slice.

CELERY : SELF-BLANCHING
Separate the sticks, scrub, trim and remove strings. Cut into short lengths and blanch 3 minutes. No use for salads, but invaluable as cooked vegetable or in cooked dishes.

CELTUCE
Peel, cut into short lengths and blanch 2 minutes.

CHICORY
Wash and blanch for 5 minutes. No good for salads, but excellent braised.

CHINESE CABBAGE
Treat as for cabbage.

CUCUMBER, RIDGE
Peel, chop, drain and freeze raw for use in recipes.

GOOD KING HENRY
Wash the leaves, blanch for 2 minutes, cool quickly and squeeze out excess moisture.

HAMBURG PARSLEY
Trim and peel. Cut into narrow strips or dice. Blanch 2 minutes. They can be treated in the same way as parsnips, or grated and eaten raw in salads.

KOHL RABI
Very young ones, no larger than 2 ins in diameter, can be trimmed, scrubbed and blanched whole for 3 minutes. Older specimens must be peeled, diced and blanched for 2 minutes.

LEEKS
Cut off green part (this can be used to flavour stock), wash thoroughly. The best way to do this is to slit down their tops 1 in or so, then stand them upside down in deep cold water for half an hour. Any dirt will drift to the bottom. Blanch whole or sliced for 3 minutes.

LETTUCE : SURPLUS HEARTS
Blanch 2 minutes, cool, then squeeze out moisture. To be used cooked, added to other soups, or for cooking peas in the French way.

MARROW AND SQUASH : COURGETTES ETC
Use only small ones and leave unpeeled. Blanch whole, or cut into $\frac{1}{2}$ in slices, for 1 minute. They can also be made into purée and packed in $\frac{1}{2}$ pint containers to use in the soufflé recipe.

ONIONS

Small onions peeled and blanched whole for 2 minutes, to use in sauce or casserole. Larger sizes can be chopped, blanched for 1 minute, for cooking later. Be sure to overwrap whatever you freeze them in, to prevent the smell escaping. As larger onions are available all the year, and are easy to store dry, it is really a question of how much space you can spare in the freezer. It is the whole, young ones, which will not keep for long, or be possible to buy later, which are well worth the space. They can also be frozen, fully cooked, in a béchamel sauce, which will only need heating-up for use.

PARSNIPS

As for Hamburg Parsley.

PEAS, GREEN

Shell and blanch at once for 1 minute.
Mange Tout: Top and tail. Blanch 2 minutes.
Petit Pois: Blanch in their pods 2 minutes. Cool and shake out peas. Discard pods.

POTATO, NEW

Blanch in skin till just undercooked to the point where there is a slight 'bite' when a cooking fork goes in. Take off skin while still warm. Let them cool in air, not water, then freeze. True pommes frittes potatoes can be cut into chips and partially cooked in deep fat for 2 minutes. Cool and freeze before final frying.

SALSIFY AND SCORZONERA

Scrub the roots and blanch 2 minutes without peeling. Cut into 2 in lengths and remove skin while still warm, then cool in air, NOT water. Or they can be cooked till almost done, then peeled, frozen, and later served in a white sauce or tossed in butter with a spot of lemon juice.

SEAKALE BEET (SWISS CHARD)

Use green part of leaf as spinach. Blanch for 2 minutes. Inner rib – cut into 2 in sticks, blanch for 3 minutes.

Above: The first sign of an onion's maturity is when the foliage starts to topple over. Help it to ripen by tucking the collapsing leaves neatly between the bulbs. *Below:* When the roots die, the bulbs can be lifted

Shallots should be lifted when their meagre onion-like tops turn yellow

French beans should be harvested when they are young enough to be stringless

Lift the swollen roots of celeriac from November onwards

Kohl Rabi must be harvested when no larger than 2 ins. in diameter. If left longer, they should be treated like turnips

Strawberries will grow in special barrels resembling colanders, punctured with planting holes

Different sorts of strawberry containers; the window box container is useful
for herbs as well

Above: The globe artichoke is often grown in herbaceous borders as well as the vegetable garden. *Below:* The Jerusalem artichoke is a hardy herbaceous perennial, resembling a potato with knobs on

By keeping the main stem and side shoots of fennel cut, or picked back, there will be a good supply of fresh leaves

When the silks at the top of the cobs of the sweet corn begin to dry and turn black, ripening has started. To test for maturity, part the sheath with the thumbs at its fattest part

Chicory ready to use after being forced in the dark

Seakale Beet is dual-purpose: the thick mid-ribs are cooked like celery and the glossy leaves as spinach

SHALLOT
As for onion.

SPINACH BEET
Wash well and blanch for 2 minutes.

SPINACH
Wash well, discard hard stems. Blanch in lots of water for 2 minutes, swishing it around. Cool rapidly and squeeze out excess water.

SWEET CORN
Pull off silks and cut away stem. Blanch for 3–6 minutes depending on size. If wanted off the cob, cool then slice off the grains with a sharp knife. This saves freezer space and makes the corn more adaptable for a variety of dishes.

TOMATO
No use for salads, but can be washed and packed whole for grilling or other cooking. Or they can be skinned, simmered in their own juice for 5 minutes, liquidised and packed as a purée.

TURNIP
Trim, peel and dice, blanch 2 minutes.

FREEZING HERBS
There are many excellent ways this can be done, depending on the purpose for which you want them.

Method 1: Chop very finely and pack tightly into ice cube compartments of refrigerator freezing trays. Top up with cold water. Freeze, and after 24 hours put the frozen cubes into labelled polythene bags. To use, take out as many cubes as you need and thaw in a strainer or sieve.

Method 2: When large quantities of a herb are needed for sauce, as in the case of parsley, mint, tarragon, they should be picked when the leaves are at their best in mid-season. Put each in turn in a blender with only just enough water to liquidise. Pour into ice trays or shallow open containers.

Turn out the blocks into polythene bags when frozen solid, and wrap individually. To use, these can be dropped straight into a sauce or stew without de-frosting.

Method 3: Wash and shake dry, then freeze in small bunches in bags or foil. To use IMMEDIATELY, you can crumble them in the palm of your hand while they are still frozen.

Note: Chives should be shredded with scissors. They turn to a mush and lose texture and juices if chopped. This is also to a lesser degree true with soft tender leaves such as basil and dill. I always use scissors to them but this must be decided by the quantities you are dealing with and their eventual uses.

FREEZING : FRUIT

The easiest and possibly the most satisfactory of all fresh frozen crops. It needs no blanching, and there are many methods, with and without sugar or syrup, which will depend on how you want to use them later.

Cautionary notes: Most freezing instructions tell us to wash *all* fruit before freezing, but with home grown produce which you know has not been sprayed or picked by anyone with unclean or suspect hands, I find this quite unnecessary. Cold water washing would be unlikely to remove anything but dust and bits and pieces, and would definitely spoil the flavour and texture of raspberries and strawberries – raspberries in particular.

I pick the berry fruits direct into their freezing containers, so they are not bruised by being tumbled from one thing to another, loosing juice and texture in the upheaval. To keep each berry individually perfect, the simplest method is to spread the dry berries on to paper-lined trays, large shallow dishes, or baking sheets. Stand them level in the quick freeze compartment of your freezer, uncovered until the fruit is frozen. They can then be moved to the main part, in whatever size of plastic bags you find most convenient. As they will each be separate it is easy enough to take or shake out, however many you want at a time. Put them in a dish to thaw at room temperature for about an hour and sprinkle with castor sugar.

APPLES

Slices: Peel, core, slice (about $\frac{1}{4}$ in thick), blanch for 2 minutes. Cool under running water. For use in pies, tarts, etc.
Purée: Peel and cook as you would for apple sauce, using the minimum amount of water. Sweet or sour, lumpy, liquidised or sieved. Cool before packing.

APRICOTS

Scald in boiling water for 30 seconds. Cool under running tap then remove skin. They can then be cut into slices, the stones removed and cut into halves, or left whole. Pack in sugar or syrup.

BLACKBERRY
BOYSENBERRY
JAPANESE WINEBERRY
LOGANBERRY
RASPBERRY
STRAWBERRY
YOUNGBERRY

Dry pack, with or without sugar

BLUEBERRIES
Wash and pack dry.

CHERRIES

Wash and dry. They can be left whole or pitted. The stones tend to give the fruit an almond flavour. They can be packed dry; spread on a tray and packed when frozen; packed dry with sugar (8 ozs to 2 lbs stoned cherries); cooked or raw and covered with cold syrup.

CURRANTS, BLACK, RED, WHITE

Top and tail, wash and drain. Dry and use dry pack method for whole fruit, or spread on tray then pack. Dry sugar pack ($\frac{1}{4}$ lb castor sugar to 1 lb fruit), mix till most of sugar is dissolved.

DAMSON

Wash. The skins toughen during freezing. Halve, remove

83

stones and store in cold syrup (1 lb sugar to 2 pints water) to be stewed later, or made into a purée for use in pies.

FIGS

Wash and dry gently to avoid bruising. Store unsweetened whole, or peeled, in polythene bags. Wrap individually in foil for dessert; alternatively peel and pack in cold syrup.

GOOSEBERRIES

Only ripe ones. Top and tail. Wash, dry pack.

NECTARINE AND PEACH

Use them fully ripe. If possible remove skin without scalding, as this will soften and slightly discolour the flesh; otherwise, put into boiling water for 30 seconds, then straight into cold, to loosen the skins. Cut into half to remove stone. Pack as halves or sliced into cold syrup. Almost half-fill the rigid containers with syrup and slice the fruit directly into them so there is no browning of the fruit. Press it well down then fill up with syrup, leaving $\frac{1}{2}$ in headspace. Dry sugar pack. Skin and slice fruit adding sugar at the rate of 1 part in volume of sugar to each 3 parts fruit slices. Mix well.

PEARS

These are tricky as they can loose texture and colour. Much will depend on the varieties you choose. Those with a strong flavour freeze the best. A case of trial and error, depending on what you grow and how you like to eat it. The fruit should be ripe but not over-ripe. Peel, core and halve, or quarter, and cover immediately with cold syrup. If you are coping with a large amount, peel and slice them into a bowl of cold water laced with lemon juice, to keep their colour till you can deal with them. They can also be poached in boiling syrup for $1\frac{1}{2}$ minutes, drained and cooled, then packed in COLD syrup. Leave headspace.

PLUMS

As for damsons.

RHUBARB

Choose young sticks and freeze early in season before it gets 'strings'. Wash, trim and cut into 1 in or more lengths. Drop into boiling water for 1 minute and cool quickly. Dry pack, or in sugar or cold syrup. It can be frozen raw – to be cooked later – but takes up more space.

WAYS WITH FRUIT

DRY PACK : UNSWEETENED

Suitable for soft fruit which can be prepared without breaking the skin, or with fruit which does not discolour while you are preparing it for the freezer. Leave ½ in headspace in containers.

DRY SUGAR PACK

This method helps draw out the oxygen from the fruit cells and is suitable for soft, juicy fruit. The quantity of sugar needed varies with the sharpness of the fruit, ranging from 4–6 ozs for each pound of fruit. They must be thoroughly mixed. This can be done in a bowl, or direct into the freezing container, putting in a little fruit then a sprinkling of sugar in layers. Leave ½ in headspace.

FREE FLOW PACK

Specially good for fruit to be used as decoration or dessert, particularly strawberries. It also enables you to take a few from a bag while still frozen, without de-frosting the whole container. Spread the dry fruit on a shallow, freezable tray, not touching. Stand them level, in the quick freezing compartment and leave uncovered until frozen. When frozen, the fruit is transferred to polythene bags in small or large quantities and stored in the main part of the freezer.

SYRUP PACK

Most suitable for hard, non-juicy fruits and those which discolour easily during preparation. The strength of the syrup depends on how sour or sweet you like any particular fruit. Make it as concentrated as possible so that it does not detract from the flavour. Be sure to have plenty made in advance as it has to be used cold. It can be stored in the refrigerator.

As a general guide, you need ½ pint of syrup (made to your taste) for every pound of fruit. Bring sugar and water to the boil, stir to prevent the sugar sticking, then cover tightly while it cools.

Syrup strength

2 cups sugar 4 cups water makes 30 per cent light syrup.
3 cups sugar 4 cups water makes 40 per cent medium syrup.
4 cups sugar 4 cups water makes heavy syrup.

A 40 per cent syrup suits most palates, but a weaker one should be used for delicately flavoured produce, otherwise you will taste more sugar than fruit.

FRUIT PURÉE

Any imperfect left-overs from fresh fruit salad and windfalls from growing crops can be cooked, as for immediate use, mashed, liquidised or sieved, then stored for future use.

CAUTIONS

Always leave headspace for fruit, however it is packed. Some fruits, such as pears, apples and peaches, start to lose their colour and turn brown as soon as they are peeled, as well as when they are de-frosting. To prevent this, you can (a) work rapidly with only small quantities, (b) slice the fruit directly into the container which has been partly filled with syrup, (c) slice the fruit into a solution of 1½ pints water to the juice of one lemon, until you are ready to pack.

Prevent fruit from floating above the syrup level of their packs by putting a piece of cellophane, tinfoil, or non-absorbent paper over the fruit and pressing down before sealing. Otherwise the 'tips of the icebergs' will turn brown.

THAWING

All frozen fruit should be defrosted in its original

UNOPENED container. Soft fruits should be served while still in a slightly icy condition to retain their shape and composure. Fruit to be used with ice-cream should be only partly defrosted. Fruit purées take twice as long to defrost as whole fruit.

Fruits to be cooked before serving need to be heated slowly from their frozen state.

Fruits wanted for open pies, etc., must be thawed until it is possible to spread them.

Fruits for use in sponge cakes, etc., should be only partly defrosted, or the sponge will go soggy.

To serve fruit frozen in syrup, leave it in its container till just before serving, to prevent discoloration, particularly with stone fruit.

PACKAGING

Inevitably, the 'learner' at deep-freezing spends so much on containers that the home produce, though better flavoured and your personal selection, costs as much as the bulk frozen packets in the local shop.

Like the first baby, it deserves the best. Yet, after only one season of freezing vegetables and fruit, you will have found that half of the packs in use are not necessary, and your own economy tricks will reduce future needless expenditure.

ORTHODOX PACKAGING

There are masses on the market, in all shapes and sizes – waxed; foil; rigid plastic-box containers; and heavy duty polythene bags. Some may be washed and used again; others, such as waxed containers, can be used again as 'moulds', lined with polythene bags, to be removed when the contents are 'set', so that the outer carton can be used again.

Whatever wrapping you use it must be :

(a) Easy to handle.

(b) Tough enough not to split or leak in the hurly-burly of finding what you are looking for.

(c) Greaseproof, moisture-vapour-proof, odourless, and capable of standing up to freezing conditions.

Much of the packaging material LOOKS as if it might be good enough, but is actually too thin for anything but sandwiches, picnic wraps, and for short storage in the domestic refrigerator. If you have to use them in an emergency, double wrap them, or, when frozen, put on an outer jacket of a guaranteed deep freeze wrapping.

ECONOMY TRICKS
Save an assortment of commercial containers for re-use, anything from cream, cottage cheese, yogurt and margarine cartons, to those made of tinfoil. If they have already been used for frozen products, you will know they have been low-temperature tested. Otherwise, to be on the safe side, line them with thin polythene. Some plastics go brittle at low temperatures and are liable to split.

Heavy duty polythene bags I have found over many years to be the cheapest, easiest and most adaptable of all containers, and can be used several times. You are able to see what is inside at a glance and to tell fruit from greens. Nevertheless labelling is still vital to differentiate between samphire, sprout tops and spinach, and the fruit berries and purées which have similar family colours.

I use my original, expensively bought, waxed containers now as moulds. The shallow rectangular variety are the most versatile. I arrange polythene bags in them, pat to shape (you don't even need the lid), quick freeze then tip from the box and stack in the cabinet in layers. Much easier than bulging bags and the wax moulds can be re-used until they collapse.

Wax tubs or plastic beakers can be used in the same way, for purées or soup. Lined with a polythene bag, wire-tied and then frozen, lifted out and stacked. Unless you have a great deal of space and some particular reason, circular tubs are a waste of space and difficult to stack, as they tumble over.

While some people swear by polythene bags, others find

tin foil suits them best. Ordinary weight kitchen foil must be used double, but heavy-duty quality can be used singly. Aluminium foil by the yard makes useful lids for cartons. Cut it large enough to fit down over the rim and seal with freezer tape or a strong rubber band.

Although it is easier to press out all the air when using foil, the disadvantage is that it punctures easily, and needs to be over-wrapped in thin polythene bags.

LABELLING

Tie clearly marked labels on to polythene bags, giving date and contents; stick special waterproof labels on packages, and write on tubs and cartons with a waterproof felt pen, or chinagraph pencil (bought at stationers). Ink smudges, and ordinary lead pencil fades in the freezer.

TIME SAVING TIPS

To make different types of food easily identifiable and save you groping head first with frozen fingers in a chest-type freezer you can use :

Different coloured (usually red, yellow, green, blue, black) sticky or tie on labels.

Lids for rigid containers.

Sealing tape.

Polythene bags.

Large storage bags for bulky items of a kind or different coloured string or nylon shopping bags for the same purpose.

PUTTING FOOD INTO FREEZER

Once you grasp the simple basic facts of freezing, you will avoid making small mistakes, which may not make the food inedible, but not leave it as pleasant in texture or full of food value as it should be.

Freezing converts the water content of food to ice crystals. Quick freezing makes tiny crystals, so that on thawing, the structure and food value are neither damaged nor changed.

Slow freezing creates large ice crystals, that damage the cell structure and produce loss of texture, colouring and flavouring when thawed.

Stick to the instructions of the manufacturer for your particular freezer. Some have special rapid freezing compartments and once the food is frozen it can be transferred to the main body of the freezer. With others, the unfrozen arrivals need to freeze against the base or sides.

Never freeze more than one-tenth of your freezer's capacity in any 24 hours.

Leave air spaces between packets of food while they are being frozen, putting new additions in contact with the refrigerated surface. Foods should reach 0°F. in 24 hours.

Keep a list of what you have stored. Cross out items as they are removed, and continue the list with new additions.

DO'S AND DON'TS REMINDERS

Blanch in small quantities, using plenty of water. If water has not returned to the boil within 1 minute, you have put in too much. Blanching time is taken from the moment the water re-boils.

Suck up surplus air through a straw when sealing bags.

Cool purées, etc., to room temperature before putting in freezer.

Freeze only top quality produce.

Harvest only as much at a time as you can deal with quickly.

Handle as little as possible, and gently – even root vegetables.

Freeze as quickly as possible, in small quantities.

Everything must be cool or cold before it goes into the freezer.

Keep everything scrupulously clean. Freezing does not kill germs or bacteria.

Never re-use polythene bags for freezing.

Don't use labels that may come off when looking for what you want.

Use only packaging material guaranteed for deep freezing – don't economise on this or you will discover contamination and unpleasant smells.

Pack food in usable quantities. It is better to take out

several small packs when you have guests than a large one of which you can only eat half. Never re-freeze remainders, or anything already thawed.

Allow air spaces between packets added to the deep freezer for rapid freezing. They can be put closer later in the main body of the machine.

Keep a record of what you put in and take out.

Frozen vegetables react like fresh produce, and the surest way of ruining their flavour and vitamin value, is overcooking. In general they should be cooked for *half* the time fresh vegetables require. They should be put, still frozen, into fast boiling water.

Exceptions: Leafy vegetables. They should be partially thawed before cooking to separate the leaves and make sure they are cooked evenly. This also applies to asparagus and broccoli.

Sweet corn. Must be completely thawed, otherwise the sheath of kernels will be cooked before the inner cob has unfrozen.

Courgettes, zucchini, etc., (baby marrows). They need never see water again. Just re-heat in butter.

Remember: Blanched vegetables are partly cooked and can be used after thawing and without re-cooking, in salads, omelettes, tossed in butter, oil and vinegar dressing or any way you fancy.

Other Ways: Steaming. Partly thaw the vegetables, to make sure the steam gets to every part.

Baking. Thaw till the vegetables can be separated and drain thoroughly. Put in greased casserole with butter and seasoning. Cover and bake in oven.

Frying. Use a heavy pan with a cover. Put in a large spoon of butter and heat to non-frizzling stage. Add vegetables while still frozen and cook gently till they separate. Cover pan and cook over moderate heat till contents are tender or as you want them. (To your personal taste.)

Frozen left-overs can be used in just the same way as cooked fresh vegetables.

Fruits that discolour must be thawed rapidly. Those frozen unsweetened must be covered in hot syrup immediately. Those to be served raw should be thawed slowly in their unopened containers, either in the refrigerator or a cool place. The berries are best eaten before completely thawed.

NEVER put them in hot water to hasten the thaw. You will end up with a liquid MESS, fit only to purée.

Those to be cooked before serving must be heated slowly from their frozen state, to retain their texture.

For open pies they must be thawed enough to spread.

For sponge cakes, etc., they should still have a bite of frost in them.

For eating raw with ice cream, only partially de-frost the fruit.

Fruit in dry sugar packs thaws more quickly than those frozen in syrup. Unsweetened fruit takes longer to thaw than sweetened. Only experience can help you to get just the results you desire. The weather, unexpected guests and a dozen other factors determine whether you will unfreeze the fruit in the airing cupboard, a sunny windowsill, an open tart, or turn it miserably into jam.

Part Three

The following recipes are mainly for home-grown frozen produce, yet they are also easily adapted for cooking fresh food from the garden, to be harvested and eaten at once, or frozen as a complete dish.

POINTS TO REMEMBER

Quantities given can be varied to suit individual requirements.

If you put unfrozen vegetables into a hot casserole or any type of composite dish, remember that they will add to the liquid content.

It is a matter of convenience whether made-up vegetable dishes are prepared from a fresh state and then frozen, so they need only be re-heated before use – as with Ratatouille; or the separate ingredients blanched and cooked together later.

Whether to de-frost or cook from the frozen state, depends on how you want your produce served. Plain boiled vegetables should remain frozen until they hit the boiling salted water.

RECIPES: VEGETABLES

ARTICHOKES: CHINESE

WITH BÉCHAMEL SAUCE

This is basically a white sauce, better flavoured. The secret of making a good, rather than indifferent one, is to cook the

95

flour and butter well before adding the milk steadily, stirring constantly as the sauce comes to the boil and thickens .You will taste the uncooked flour if the sauce is made in a hurry.

For 4 people. 1 oz butter; 1 oz flour; ¼ pint milk (for binding consistency); ½ pint (for sauce or coating consistency); 1 pint (for thin sauce and soups); small piece of carrot, onion, celery, and bayleaf or other herbs if desired.

Heat the milk with vegetables to just short of boiling. Cover to retain flavour and leave in warm place for about an hour.

Melt the butter gently, stir in flour and cook for a few minutes. Gradually blend in the strained milk, bring to the boil and stir with wooden spoon until smooth, simmer very gently for a minimum of 10 minutes, stirring frequently. There should not be lumps if the initial stages are carried out properly, but if things go wrong, press the sauce through a fine sieve into a clear saucepan. An electric hand beater or liquidiser may rescue you.

Add the slightly undercooked artichokes to the hot sauce in the proportion of ½ pint of sauce to 1 lb of artichokes.

CROQUETTES

Make a thick Béchamel sauce, as above, and bind with 2 well-beaten egg yolks added to mixture while warm. Cool, shape into croquettes, brush with melted butter, coat with breadcrumbs and fry in deep fat.

PURÉE

To every lb of artichokes add ¼ lb potatoes diced small. Cook in just enough water to cover; sieve or liquidise with a little milk, re-heat in saucepan adding enough milk and butter to make a light purée, seasoned with salt and pepper.

SAUTÉ

Boil till artichokes are cooked but still crisp. Drain thoroughly and cook in butter in heavy pan. When lightly browned, season and sprinkle with a little chopped parsley and lemon juice.

WITH VINAIGRETTE SAUCE

Cook the globes until the outer leaves can be pulled away. Drain upside down and serve when cold with the sauce. They are eaten by dipping the base of each leaf into the sauce, eating the soft part (in one bite), discarding the rest. When the heart is reached, the barb-edged 'choke' is removed, and the heart, which is actually the tender base, is eaten sprinkled with or dipped into the sauce.

VINAIGRETTE SAUCE: (FRENCH DRESSING)

1 tablespoon vinegar or lemon juice to 3 tablespoons salad oil, 2 teaspoons sugar, 1 teaspoon mustard, salt and pepper to taste.

VINAIGRETTE WITH SOFT OR HARD BOILED EGG

Mix 3 tablespoons olive oil and a little lemon juice or wine vinegar with a finely chopped shallot. Add salt and pepper, a tablespoon of parsley and a few chives cut with scissors (chopping turns them into a mush). Boil an egg for 3 minutes, stir the scooped-out yolk into the sauce and add the chopped white.

A hard boiled egg can be used instead and a few gherkins added.

Alternative: Melted butter mixed with black pepper, lemon juice and sugar. Proportions are a matter of taste. Hollandaise, or Mousseline sauce, or mayonnaise.

ARTICHOKE HEARTS WITH MUSHROOMS

4 globe artichokes, 8 ozs mushrooms, 2 tablespoons butter, 1 tablespoon grated cheese, preferably Gruyère.

Boil, drain, peel off all leaves and remove choke being careful not to crack the bottoms. Melt butter in frying pan and gently fry the bottoms. Remove and keep warm, then fry mushrooms, finely sliced in the same butter. Fill artichoke bottoms with the mushrooms, give a generous sprinkle of grated cheese and heat under grill till the cheese melts.

FRICASSÉE

Remove the coarsest outer leaves, and cut away more than half of the pointed end. Place in thick saucepan with a little butter, chopped parsley, clove of crushed garlic, and salt. Cook slowly and, when tender, remove the pan from heat; add one or two egg yolks (depending on the quantity of artichokes) beaten lightly with a teaspoon of water per yolk, the juice of $\frac{1}{2}$ a lemon and a tablespoon of grated Parmesan cheese. Serve sizzling hot.

WITH ANCHOVIES

Remove outer leaves and half of the top; cut into quarters downwards. Cook in a little oil, lemon juice, crushed clove of garlic, salt, pepper and a pinch of oregano for about 15 minutes, turning them frequently so that they become golden all over. Remove artichoke quarters to hot dish and add one or two anchovies, cut in small pieces, to the sauce. Cook for a few minutes then pour over the artichokes.

FRIED ARTICHOKES

4 small globe artichokes; 1 egg; 4 ozs flour; 1 teaspoon olive oil, $\frac{1}{2}$ pint water.

Cook till tender. Slice in half and remove chokes. Then cut each half into pieces and drain well. Make a batter by beating the egg thoroughly adding the flour gradually, then the water, salt and oil. When creamy, dip in each piece of well drained artichoke and fry in very hot deep fat. Once the batter is golden, remove, sprinkle with salt, and eat at once.

STUFFED ARTICHOKES

4 medium sized globe artichokes; 8 ozs breadcrumbs; 4 ozs sausage meat, 4 tablespoons stock; 1 small onion; salt and pepper; 2 ozs chopped parsley; 4 ozs mushrooms; 2 ozs butter; 1 tablespoon flour; 1 tablespoon olive oil. 2 lemons.

Cook artichokes and drain well. Pull off tops to use later. Cut out chokes carefully. In a bowl, mix breadcrumbs with stock, sausage meat, mushrooms cut into small pieces, chopped onion and parsley. Melt butter in frying pan till sizzling hot; fry the sausage meat in it and once it becomes golden, blend in the flour. Fill the centre of each artichoke with stuffing, replace tops, and put into well-oiled ovenproof dish. Bake in hot oven for 30 minutes and serve garnished with lemon slices.

À LA PROVENÇALE

Use 6 small artichokes. Put in an earthenware casserole with about $\frac{1}{4}$ in hot oil; season with salt and pepper; cover and simmer for 10 minutes. Then add $\frac{1}{2}$ pint small peas and a lettuce, coarsely shredded. Cover and simmer for another 30–45 minutes, or until artichokes and peas are tender. A pinch of sugar can be added with the peas.

STUFFED, ITALIAN STYLE

Prepare 6 medium to large artichokes by removing outside leaves, cutting off about 1 in of the top of the leaves and scooping out the choke.

Make stuffing with 2 tablespoons breadcrumbs, 4 chopped anchovy fillets, and 2 chopped cloves of garlic. Cover the bottom of a small deep pan with olive oil, and when warm put in stuffed artichokes. Add 1 glass of white wine and simmer very gently for about an hour. Serve hot.

WITH HAM

Cook, and when tender, remove leaves, hearts and chokes. Sprinkle bottoms with lemon juice and fill with a mixture of finely chopped ham and hard-boiled egg bound with mayonnaise. Serve chilled.

ARTICHOKES AND BROAD BEANS : (GREEK)

Cook separately 2 lbs broad beans and 8 artichoke bottoms. Strain and keep $\frac{1}{2}$ a cup of the bean water. Heat 2 tablespoons olive oil in a pan, stir in a very small amount of cornflour, the liquid from the beans, the juice of a lemon, chopped parsley, then the vegetables, and heat them through.

99

SOUP À LA PROVENÇALE

Cook 2 lbs peeled artichokes in 3 pints salted water. Sieve or liquidise and re-heat gradually, adding ½ pint milk (or this can be added when liquidising).

Heat 2 tablespoons olive oil in small frying pan; add 2 chopped tomatoes, a clove of garlic, small piece chopped celery, a little parsley, and 2 chopped tablespoons of ham or bacon. Let the mixture cook only a minute or two so it tastes fresh and not stewed, then pour it with the oil into the soup. Heat and serve quickly.

STEW

Cut 2 lbs artichokes into quarters. Melt 2 tablespoons butter in large saucepan. Finely slice a large onion into it until it begins to brown; add artichokes; sprinkle with salt, pepper (a little sugar if liked), a crushed clove of garlic, a little nutmeg and a bouquet garni. Cover with 1½ pints stock, a gill of white wine, cover and simmer gently for 20 minutes.

AU GRATIN

Butter a baking dish; add 4 cups boiled and diced artichokes; pour over the mushroom sauce (Béchamel with sliced mushrooms added; see Chinese artichokes), and sprinkle with 6 ozs breadcrumbs fried in 2 ozs melted butter. Put in hot oven for 3 minutes.

PURÉE

Make a thin roux (a simpler version of a Béchamel) by melting 1 oz butter, adding 1 oz flour; cook a few minutes, and gradually add 2 pints warm milk or 1 pint milk and 1 pint stock. Stir till smooth. Add 12 ozs puréed artichokes (take from the 2 pints milk if using a liquidiser). Cook gently for 4 minutes and stir in 1 oz of grated cheese, preferably Parmesan. Serve extra grated cheese in a separate bowl.

SAUTÉ

Dry 1 lb under-cooked sliced artichokes. Heat 1 tablespoon

each of vegetable oil and water. Put in artichokes, raise heat for a few moments after putting on lid, then reduce it and cook gently till the artichokes still have a 'bite' to them. They toughen if cooked too fast.

FRIED
Roll partly cooked artichokes in breadcrumbs and deep-fry.

GRILLED
Cut blanched, de-frosted artichokes into fingers and toss in a bowl with 1 tablespoon of oil. Put in pan under grill at high heat for about 3 minutes, when they should have browned. Turn with slicer to brown other side at lower heat for about 5 minutes more.

ASPARAGUS

Thick, high quality stems are ruined by anything but the simplest treatment. Cook gently (the tips should have less heat than the thick ends). This can be done by standing the bases in water and leaving the tops to steam, or using a long shallow pan of water, laying the stems lengthways, with only the thick ends at boiling point, and tips off the heat.

Serve with melted butter, Hollandaise or vinaigrette sauce.

SOUP
This can be made in endless ways, depending entirely on what quality of 'grass' you are willing to sacrifice. If you use woody, left-over parts, these are impossible to sieve and can only be strained, leaving a wish-washy taste, which will need additional flavouring and thickening.

Basic Sunday Best recipe. Cut tender stems into 1 in lengths, keeping heads separate. Heat in butter for 10 minutes, then stir in enough flour to absorb the butter, add liquid – water, milk, stock – simmer till tender, then liquidise or sieve.

Return to pan and add cream, or more of your chosen liquid, seasoning, and at the last moment, the asparagus tips, gently cooked in butter.

SPREW (THIN GREEN ASPARAGUS) MOUSSE

Ingredients.

¾ lb sprew; milk salt and pepper; 1 oz flour; 1 oz grated Parmesan cheese; 3 separated egg yolks; 2 tablespoons cheap dry white wine; ¼ oz gelatine; 2 egg whites.

Chop or cut sprew into small pieces, put in tallish sauce pan and add milk just to cover. Simmer very gently till tender. Strain into another pan or bowl, keeping the residue milk, and rub the soft sprew through a sieve.

Whip egg yolks with the flour to a smooth paste. Measure the residue milk and make it up to half a pint with top milk or thin cream. Heat milk and when boiling pour on to the egg yolk batter. Blend thoroughly, pour into top of double saucepan and stir until thick and smooth.

Melt gelatine over a thread of heat with the wine. Stir into the savoury egg mixture with the cheese and sieved sprew. Taste and correct seasoning, then fold in the stiffly whipped egg whites.

Put into individual containers, or one large one, and serve well chilled, with hot toast, or as a filling for avacado pears.

Other uses: Sprew asparagus, re-heated with melted butter and served with scrambled eggs, hot or cold.

Asparagus tips as garnish for anything, from open savoury tart, to shrimp vol au vent.

Cold tips sprinkled with mashed egg yolks and thin strips egg white, then dressed with vinaigrette and chopped parsley.

ASPARAGUS, PEA

Subtle taste destroyed if anything but butter is added. Serve on its own, steamed or boiled till crisp but not soggy.

BEANS, BROAD

WITH PARSLEY SAUCE

Cook and strain beans and add to plain white roux, Béchamel sauce, or thick cream with fresh parsley. Garlic salt or fresh pounded garlic makes an interesting difference.

WITH CHOPPED HAM

Prepare any favourite white sauce (as above), add beans, and at last moment the ham. Be careful not to oversalt the sauce.

STEWED IN OIL

Unthaw broad beans (under tap if pressed for time), put into heavy saucepan with about 1 in of oil. Cover and cook, shaking pan occasionally for about 20 minutes. Then add a topping of ripe skinned tomatoes, seasoned with sugar, pepper and salt and fresh basil. Cook for another 10 minutes. It can be eaten hot or cold, but it is best just slightly warm.

WITH YOGURT

2 lbs of cooked broad beans mixed with 2 tablespoons cooked rice. Heat slowly with 5 ozs yogurt plus a scrap of garlic. Stir in a well beaten egg and chopped parsley or mint, or any other favourite herb.

BEANS, FRENCH, DWARF AND RUNNERS

At their purest, steamed or boiled whole, or snapped into pieces, till tender but still crisp. Served hot with melted butter, or cold in vinaigrette.

WITH MORNAY SAUCE

To every pint of Béchamel sauce add a sparse 1 oz each of finely grated Gruyère and Parmesan cheese. Stir over a slow heat till the cheese melts, and before serving add 1 oz butter, divided in small pieces. Pour on to the hot cooked and drained beans.

À LA CASTELLANA : (SPANISH)

Cook beans till tender. Cut up and fry in oil, sweet red peppers, a garlic clove and chopped parsley. Add strained beans to mixture. Proportions of peppers and beans are a matter of one's individual palate.

RUSSIAN STYLE

Cook 1½ lbs broken or sliced French beans. Slice and fry 8 ozs mushrooms lightly in 2 tablespoons butter. Add 2 tablespoons of water in which beans were cooked and simmer for 5 minutes. Add beans, stir in a gill of sour cream, heat for another few minutes and serve sprinkled with chopped parsley.

À LA LIONESA

Cook and dry 2 lbs beans. Chop up 4 ozs smoked ham with 2 small onions and fry gently in 3 ozs butter till golden. Add beans and heat, shaking pan so they do not stick. Sprinkle with parsley or any favourite herb.

FRENCH BEANS WITH PEARS : (GERMAN)

In salted water, partly cook 1½ lbs broken beans; add 1 lb peeled, cored and quartered pears (not too juicy a variety), and continue to cook till both are tender.

Make a sauce with 2 ozs butter, 1 tablespoon flour, and ¾ pint of stock from beans and pears. When smooth and thickened, flavour with juice of ½ a lemon. Add sugar and salt to taste. Re-heat beans and pears and serve at once. This is particularly good with pork.

SAUTÉ

1 lb whole, young stringless beans, boiled strained and added to a small sliced onion cooked till soft in 3 tablespoons butter in large frying pan. Keep tossing and when beans begin to brown, season with plenty of black pepper.

SWEET-SOUR GREEN BEANS : (AUSTRIAN)

Bring 1½ gills water to boil in large saucepan; add small

sliced onion, salt, pepper, bayleaf, clove of garlic and a little nutmeg. Let these simmer for 10–15 minutes, then add 1 lb beans. Cover and boil for 10 minutes. Strain into another pan and keep beans warm after fishing out garlic and bay leaf. Put back liquid; add 1½ tablespoons sugar, 1½ tablespoons tarragon vinegar, 2 cloves and stir in 2 tablespoons butter, then add beans and serve with chopped parsley.

GREEN BEANS WITH FENNEL

Use two separate saucepans with only small amounts of boiling salted water. Put 1 lb beans into one and 4 ozs chopped fennel in the other. Cover both tightly and cook for 10 minutes, shaking the pans occasionally so the contents do not stick. Remove from heat, drain off water and put beans in a warmed dish with the fennel strewn on top. Keep warm while you make a butter sauce.

Melt 1 oz butter with 1 tablespoon flour, blend well. Gradually add 1 gill water, stirring constantly until sauce thickens. Add another 1 oz butter, 1 teaspoon lemon juice, a grating of nutmeg and 1 tablespoon grated onion. Stir all the time and don't let the sauce boil or it will curdle. Pour over the dish of beans and fennel and serve at once.

GREEN BEAN AND TOMATO SALAD

Cut up 1 lb of undercooked beans with 8 ozs peeled and quartered tomatoes. Toss them in dressing of 3 tablespoons olive oil, 1 tablespoon white wine vinegar, small clove, crushed garlic, salt, pepper, a scrape of sugar and chopped basil.

RUNNER BEANS WITH EGGS : (BULGARIAN)

Chop two onions and fry lightly in 2 tablespoons of oil. Add 1 lb sliced runner beans. In separate pan, brown a teaspoon of flour in a tablespoon oil, add ½ teaspoon paprika and a tablespoon chopped parsley. Combine with beans and onions, mix well and put into fireproof dish. Beat 2–3 eggs with ½ pint of milk, add pinch of salt and pepper. Pour over bean mixture, cover and cook in moderate oven for 15 minutes.

SOUP, ICED

Dice 2 raw or blanched beetroots, a carrot and onion. Bring to boil. Add 2 cups stock and boil for 10 minutes until tender enough to sieve or liquidise; add seasoning and put in fridge. An hour before serving, shred a cucumber, cover with salt and put in bowl in fridge. To serve, squeeze liquid out of cucumber, put in tureen and cover with iced beetroot. At last moment swirl a carton of yogurt or sour cream on to the surface.

AS SAUCE

Mince or grate 2 lbs beetroot finely. Put into saucepan with 2 tablespoons cream, juice of $\frac{1}{2}$ lemon, pinch of salt. Heat for 5 minutes and serve with meat dishes.

BEETROOT AND HORSERADISH SAUCE

Melt 2 tablespoons butter, blend in a tablespoon flour, dilute with 1 pint stock and bring to boil. Add 8 ozs grated horseradish, 1 teaspoon dry mustard, 1 gill vinegar, 1 teaspoon each of salt and pepper, and simmer for 10 minutes stirring constantly. Add 4 ozs cooked grated beetroot and heat another 5 minutes. Serve cold with aspic dishes, cold beef, pork and fish.

BROCCOLI : SPROUTING

Top quality spears (heads) are at their best served like asparagus – hot, dipped into melted butter; cold dipped into lemon and oil, or with any of the butter or oil-based sauces.

SAUTÉ

Heat butter, oil, or a mixture of both, in thick pan. Sauté finely chopped shallots until transparent; add thawed broccoli stems, cut into small pieces; stir well. Add 1 tablespoon water, raise heat, cover and cook a few minutes, then add the broccoli buds (which will not need as much cooking as the stems) and cook for 8–10 minutes or until they are at the stage of 'done-ness' you prefer. Two peeled and chopped ripe tomatoes can be added for the last 2 minutes, but don't turn the dish into a stew by adding too much liquid – only enough for the broccoli to absorb and cook in without burning.

WITH HOLLANDAISE SAUCE: (QUICK METHOD)
Slowly melt 4 ozs butter in heavy pan. Remove as soon as the first bubbles appear. Heat container of liquidiser, in oven or very hot water. Put in 3 egg yolks and 1½ tablespoons of lemon juice and blend. When well mixed start to add the melted butter a trickle at a time as you would for mayonnaise. Leave behind the white sediment at the bottom. Continue to blend until sauce is thick and velvety. To re-heat, pour into double saucepan, or a bowl resting over, but not in, simmering water. Stir till sauce is hot, season to taste and spoon at once over cooked, hot, drained broccoli.
Simplified traditional method: Use double saucepan or bowl over simmering (but never boiling) water. Put into top, 2 egg yolks, salt and pepper, 1–2 tablespoons lemon juice or white wine vinegar. Whisk till sauce starts to thicken. Then add 2–4 ozs butter cut into very small pieces, whisking in each piece until completely melted before adding the next. It will curdle if you allow it to boil.

BROCCOLI WITH CHEESE AND ALMONDS
Make a white sauce (See Béchamel Page 95). Add 3–4 ozs grated cheese (a mixture of Emmenthal and Parmesan, or any you prefer). Stir occasionally till cheeses have completely melted. Meanwhile, cook thawed broccoli spears in boiling salted water till tender, and toss 2 ozs of blanched and slivered almonds in a little hot oil until golden. (Beware, they suddenly start to burn.) To serve, put drained broccoli into ovenproof dish, cover with sauce, sprinkle with almonds and leave in oven until wanted, or brown lightly under a hot grill.

WITH CHESTNUTS

Boil 1 lb small sprouts in salted water till just cooked and still very firm. Drain well, then lightly brown in 1 oz of butter. Stir in half the quantity of chestnuts previously cooked and peeled, plus a teaspoon of sugar.

This dish can be varied by adding 4 ozs of crisply fried strips of bacon.

PURÉE

Boil sprouts for 5 minutes; drain and cook in butter till tender; sieve. Allow one part of potato purée to two of sprouts. Mix together and beat well, over low heat. Add seasonings, grated nutmeg, and a little hot milk or a few lumps of butter to suit your taste and the texture of the purée.

WITH CREAM

Boil 1 lb baby sprouts to a point where there is still a 'bite' in them. Drain thoroughly; they must be quite dry when added to the cream. Boil $\frac{1}{4}$ pint single cream until it is reduced to about half; add the sprouts, a nut of butter, salt and pepper. Cover and shake pan to mix contents thoroughly. Leave on a low heat for a few minutes to allow the flavours to get together.

WITH CROÛTONS

Boil 1 lb sprouts. Drain, press, chop and set aside. Chop a shallot; crush a clove of garlic in salt and cook gently in 1 oz of butter till lightly brown. Add the sprouts, seasoning, and a squeeze of lemon juice. Serve garnished with fried croûtons.

WITH CELERIAC OR CELERY

Cook 1 lb sprouts; parboil a root of celeriac or a stick of celery and chop. Soften a medium chopped onion and the celeriac in $1\frac{1}{2}$ ozs butter for a few minutes, then sprinkle in 1 oz flour; dilute with $\frac{3}{4}$ pint or less of milk. Cook for a few minutes so the flour loses its 'raw' taste, then add cooked sprouts, and turn mixture into fireproof dish. Sprinkle with a few breadcrumbs melted in butter and finish off in a hot oven.

Alternatives: Sauté, or with any egg or flour-based sauce with a flavour not strong enough to destroy that of the sprouts.

CABBAGE

WITH BABY ONIONS
Boil shredded cabbage rapidly in just enough water to cover, with about half the quantity of small whole onions. Drain, toss in butter, and sprinkle generously with minced parsley.

WITH SOUR CREAM
Put shredded cabbage into deep frying pan with 1 tablespoon butter and 3–4 tablespoons sour cream. Season highly with salt and a little pepper. Cover pan and simmer very gently for an hour or more, stirring occasionally. The cabbage should be very tender, similar to a mild sauerkraut.

WITH CHEESE
Put a layer of blanched cabbage leaves in a buttered fireproof dish. Cover with a layer of thinly sliced Gruyère cheese, and continue filling dish with alternate layers of cabbage and cheese, ending with cheese. Sprinkle with breadcrumbs; moisten with melted butter and brown in a quick oven.

This dish can be varied by adding chopped, cooked ham between the layers.

SWEET-SOUR CABBAGE
Heat a little olive oil in large saucepan and sauté a small onion. When golden, add 2–3 large, ripe, peeled tomatoes. When soft, add shredded cabbage, salt, pepper and a tablespoon of wine vinegar (NOT malt). Stir and let simmer for 20 minutes. 5 minutes before serving, stir in 1 tablespoon of castor sugar.

Use partly-cooked leaves and on each put a small spoonful
of what you fancy – sausage meat, forcemeat, savoury rice,
mushrooms, bacon and rice – roll up the leaves neatly;
squeeze in the hand to keep them firmly packaged. Put them
closely into a buttered fireproof dish, moisten with butter
and brown in a moderate oven.

Variations: The stuffed leaves can be closely packed in a
saucepan on a layer of bacon rashers, diced carrots and
onions, and moistened with a little stock. Cover and simmer
for an hour. Or they can be braised, covered, in oil or stock.
A happy way of using up cold meats.

WITH MINCED MEAT
Chop and fry an onion in a tablespoon of butter. Add 8
ozs minced meat and seasoning and fry for a few minutes.
Beat 2–3 eggs in 1¼ pints milk; add a pinch of sugar, pour
over the meat and mix well. Meanwhile, partly cook sliced
cabbage; drain and lay in a casserole. Pour over it the meat
mixture; sprinkle the top with breadcrumbs and bake for an
hour in a moderate oven.

CALABRESE

They need very little re-cooking and can be served on their
own and eaten like asparagus, with melted butter or any of
the accompanying sauces. They can also be cooked in any
of the ways suitable for the broccoli family. Over-cooking
ruins the delicate flavour.

CARROT

NEW CARROT SOUP: CHILLED
Cut up 1 lb new carrots with 3 new potatoes and 1 doz little
silver onions. Put in saucepan; just cover with water; add

salt and a few sprigs of parsley, then cook till tender. Sieve or liquidise; add 1½ pints rich milk; ¼ pint cream, and ½ glass white wine. Mix thoroughly and chill. It should be a lovely golden colour, flaked with green from the parsley. More fresh parsley can be added at the last moment.

ICED CARROT AND ORANGE SOUP
Scrub and slice thinly 1 lb new carrots; slice an onion and fry them gently in 1 oz butter till soft, but not brown. Stir in 1 pint mild stock (or water and chicken stock cube), salt and 1 teaspoon of sugar. Bring to boil; cover and simmer gently until soft. Sieve or purée in blender; add the strained juice of 4 oranges. Chill for several hours before serving.

VICHY
Slice 1 lb of new carrots diagonally, ¼ in thick. Put into heavy pan with 2 ozs butter, 1 oz sugar, a pinch of salt, and ¾ pint of water. Cook uncovered until water is almost evaporated and carrots tender. Add another lump of butter and shake to prevent carrots from sticking. Add chopped parsley before serving.

GLAZED
Simmer 1 lb of carrots, sliced or in strips in 4 level tablespoons butter, 4 tablespoons chicken stock, and 1 level tablespoon of sugar and salt to taste, until carrots have absorbed the liquid without burning and are slightly coloured.
Variations:
Lemon-glazed. Cook as above but include the grated rind of 1 lemon and a tablespoon of juice at start of cooking.
With Béchamel. Cook as above, then just before serving add 4–6 tablespoons Béchamel sauce.
With herbs. As above with sprinkle of 2 tablespoons each of chopped parsley and chervil.
Oriental. As above, adding 2 tablespoons pre-soaked raisins at the same time as butter and chicken stock.

RING MOULD WITH PEAS AND ONIONS
Cook as for glazed carrots. Mash and mix well with an egg,

2 tablespoons softened butter, 3–4 tablespoons freshly grated cheese, salt and black pepper to taste. Press into well-buttered ring mould and heat in moderate oven for 15 minutes. Turn out on to hot serving dish and fill centre with cooked peas and button onions.

CELERIAC

STEWED IN BUTTER
Cook shredded celeriac in butter in a frying-pan for 10 minutes, turning over and over. Before it is quite ready, add salt, pepper, a teaspoon of French mustard, a dash of tarragon or wine vinegar and a little chopped parsley. The celeriac should remain slightly crisp and not be cooked until soggy.

BRAISED
Line a casserole with a few strips of bacon, sliced onions, carrot, and a bouquet of herbs. Add sliced celeriac. Pour over stock to half way up vegetables; bring to boil; cover and simmer in oven for about 1 hour and serve when tender. *Alternatives:* Celeriac can be used in any salads and stews, or served on its own with sauce.

CELERY

FRITTERS
Boil until tender, then divide into $\frac{1}{2}$ in rounds. Dry, season, and dip 1 tablespoon at a time in frying batter, and cook in deep fat. Drain well on paper before serving.

Frying batter: Put 3 ozs of flour into large bowl. Make a well in middle and put in 1 tablespoon oil, 1 egg yolk, salt and pepper. Gradually mix in flour and a gill of hot milk, little by little, until batter is smooth. Cover and leave in a

warm place for an hour or so. Add two stiffly beaten whites
of egg just before using.
Alternatives: Braised; as for celeriac. Boiled and dressed
with Béchamel or Mornay sauce. Frozen celery is at its best
in hot cooked dishes rather than salads.

CELTUCE

Use as you would celery, cooked or raw. Blanched or raw
¾ in sliced rounds can be used as bases for cocktail snippets,
or cut into strips and dipped into various cold 'dip' sauces.

CHICORY

BRAISED WITH WALNUTS
Heat 1 tablespoon of butter in large saucepan, add scissored
fresh basil or ¼ teaspoon of it dried. Put in 4 heads of chicory
sliced in half lengthways and brown for 2 minutes on each
side. Add 1 cup of stock, a little at a time, and simmer until
chicory is tender, adding more stock as required. Serve
garnished with walnuts fried in butter.

AU GRATIN
Boil 4 large heads. Drain and arrange on greased fireproof
dish. Make a sauce with 2 ozs butter, 2 ozs flour, ½ pint
milk, and 2 ozs grated cheddar cheese. Whisk sifted flour into
the hot milk and butter and stir vigorously on low heat till
the sauce thickens. Then add cheddar cheese and stir until
this has melted; pour mixture over the chicory, sprinkle on
1 oz grated parmesan cheese and cook in a hot oven till the
sauce is slightly browned on top.
Alternative au Gratin: Prepare as above, but wrap each
head in a thin slice of ham before covering with sauce, and
sprinkle with freshly made breadcrumbs and melted butter.

Cook in a moderate oven until the sauce is bubbling and golden.

BAKED WITH LEMON

Put 6 heads of chicory into large, preferably glass, oven dish, with 1 tablespoon butter, 4 tablespoons fresh lemon juice, and 1 tablespoon sugar. Cover with buttered paper and cook in oven till chicory is tender and golden. Look occasionally, and if lemon juice has dried up, add a little water.

CHINESE CABBAGE

Cook in any way you would treat lettuce or cabbage.

CUCUMBER : RIDGE

LEBANESE COLD CUCUMBER SOUP

Three cups peeled cucumbers sliced paper thin; $\frac{1}{4}$ cup chopped fresh mint; a clove or crushed garlic; $\frac{1}{2}$ teaspoon coarse black pepper; salt to taste, 3 cups yogurt.

Mix the yogurt, cucumbers, the seasoning and half the mint. Refrigerate and just before serving mix in crushed ice (if needed) and sprinkle the rest of the mint on top.

ALTERNATIVE ICED CUCUMBER SOUPS

(a) 1 large unpeeled cucumber, 2 cartons of yogurt, crushed clove of garlic, 1–2 tablespoons tarragon vinegar, salt and pepper, 1 tablespoon finely chopped cocktail gherkins, 2 ozs peeled prawns. (You can leave out anything that cuts across your particular taste grain.)

Grate the cucumber, stir the pulp into the yogurt, add the seasonings and chill. Just before serving, stir in the gherkins and prawns.

(b) 1 large cucumber, 2 shallots, 2 tablespoons butter, 1 desertspoon flour, 1$\frac{1}{2}$ pints chicken stock, 3 egg yolks, $\frac{1}{2}$ pint single cream, pepper and lemon juice, with chopped mint and diced cucumber to garnish.

Cook peeled and chopped shallots and cucumber in butter till soft, stir in flour, pour over stock, season, cover and simmer about 20 minutes. Add lemon juice to taste. Sieve or blend and return to pan. Beat egg yolks with the cream, blend in a few spoons of the hot soup, then return to the soup in the pan and stir till it thickens. It must not boil. Cool then chill. Serve garnished with finely chopped mint, small dices of raw cucumber and some prawns floating on top.

(c) 2 medium sized young ridge cucumbers, $\frac{1}{2}$ pint of plain yogurt, $\frac{1}{4}$ pint of soured cream, a large clove of crushed garlic, 1 desertspoon of lemon juice, 1 teaspoon fresh chopped mint, salt and black pepper.

Peel and slice the cucumbers thinly. Keep some aside as a garnish and put the rest in a liquidiser. Add yogurt, soured cream and garlic. Blend till smooth. Pour into a bowl and thin it slightly with a little milk if it seems too thick. Stir in the lemon juice, seasoning and mint, cover and chill thoroughly. Serve topped with the remaining cucumber slices. *Alternatives:* Cucumber can be peeled, cut into 2 in lengths, boiled until tender but not squashy. Drain well and serve with any favourite sauce.

STUFFED CUCUMBER

Peel cucumbers, and cut into 2 in lengths. Blanch in boiling salted water for 2 minutes, then toss at once into cold water. Drain and scoop out seeds and part of pulp. Stuff with veal forcemeat, sausage meat, or anything your palate can take. Put on layer of bacon rashers in saucepan; add enough stock barely to cover, put on lid and simmer gently till cucumbers are tender.

FRIED CUCUMBER

Peel and slice 1 large cucumber and 2 ozs spring onions and fry in 1 tablespoon of hot butter till soft. Stir in $\frac{1}{4}$ pint yogurt or sour cream; simmer for 2 minutes. Serve with hot meat.

HAMBURG PARSLEY

WITH CELERY AND EGGS

Steam or boil in a small amount of water for about 30 minutes, then add a head of celery cut into lengths. Continue until both are tender. Drain, put into buttered casserole dish, pour over some cream or white sauce, and break one egg for each person on top. Cook in moderate oven till eggs are set, and sprinkle with chopped parsley before serving.

Alternative uses: Hamburg parsley can be used for any recipe given for parsnips. Raw, it can be grated into salads or used on its own. Cut into cubes, it can be speared on to cocktail sticks with cheese, prawns, slices of frankfurter, or any mixture that appeals to you.

KOHL RABI

They can be treated like turnips. Or the still slightly frozen slices can be used as 'bases' instead of bread or biscuits for cocktail savouries. They can also be cooked, sliced and fried in egg and breadcrumbs as fritters.

LEEKS

VICHYSSOISE

Toss 8 medium sized leeks in 2 ozs melted butter till pale gold. Add 1½ pints chicken or white stock, 2 medium sized chopped potatoes, 1 tablespoon chopped parsley and 1 tablespoon chopped chives. Season and simmer gently for 30 minutes.

It should then be sieved, but you can cheat with a liquidiser as long as the mixture is well thinned down with stock. When cool, gradually add ¼ pint of cream; chill and serve topped with more chopped chives.

A LA GREQUE

You need small leeks, and this is a way of saving those in the seedbed which were not needed for planting out. Boil till tender in only a little water and leave them in it to get cool. (There should be just enough to cover them.) Stir some of the liquid into a bowl containing 1 teaspoon of cornflour; add to the leeks and stir until the sauce thickens a little on a gentle heat. Squeeze in the juice of a lemon, stirring all the time and add 1 tablespoon of olive oil. Let the leeks cool in the sauce, and serve as a beginning dish or salad.

TART

Line a well buttered tart or flan tin with a thick layer of shortcrust. Simmer cooked, chopped leeks, in a little butter, and 1 or 2 well-beaten egg yolks; put into tin and cover with another layer of pastry. Cook in a hot oven till the pastry is golden. Serve very hot.
Variations: Add any cooked chopped cold meat or poultry, or cheese to the filling.

SOUFFLÉ

Slice, cook and drain 8 good sized leeks. Return to pan and dry off in $\frac{1}{2}$ oz butter. In another saucepan melt 1 oz butter, add $1\frac{1}{2}$ ozs flour, cook slightly and gradually stir in $\frac{1}{2}$ pint warmed milk. Stir till boiling. Remove from heat, add salt, pepper and nutmeg, 4 egg yolks and beat well. Pour on to leeks and mix gently. Whip whites till stiff, mix in a large spoonful and fold in the remainder. Turn into a prepared soufflé dish, dust with breadcrumbs browned in butter and bake in moderately hot oven for 20 minutes, or until it has risen but is still slightly moist inside when tested with a skewer.

PROVENÇALE

Heat 1 tablespoon of oil in a shallow heatproof dish. Add 1 lb thawed leeks cut into $\frac{1}{2}$ in lengths, cover and simmer for 10 minutes. Add $\frac{3}{4}$ lb halved tomatoes, 6 stoned black olives, a sprinkle of lemon juice and the grated rind of $\frac{1}{2}$ a lemon. Cook slowly for a further 10 minutes. Serve in the dish in which it was cooked, either hot or as a salad.

LETTUCE

SOUP

Cut 2 lettuce hearts into strips with $\frac{1}{2}$ the amount of spinach, slice 6 spring onions, simmer in 2 tablespoons butter for 10 minutes, stirring occasionally. Add 1 pint chicken stock, bring to boil, reduce heat, cover and simmer for 30 minutes. Sieve or liquidise. Return to saucepan, stir in $\frac{1}{2}$ pint Béchamel sauce until boiling, then add $\frac{1}{4}$ pint cream and finely chopped chervil or parsley. Serve with croûtons.

WITH PEAS FRENCH STYLE

Put a layer of shredded lettuce (it need not be thawed) into a heavy pan. Add 8 small whole onions, $1\frac{1}{2}$ lbs peas, 1 teaspoon sugar, 2 ozs butter and 2 tablespoons water. Cover with more lettuce, replace lid and cook steadily until the peas are tender. (You may need more or less water depending on the dampness of the lettuce.)

BRAISED

Line a saucepan with a few rashers of bacon, 2 ozs each of sliced carrot and onions and a bouquet of herbs; add 6–8 whole blanched lettuces closely packed, cover and simmer gently and when lettuces are lightly browned, barely cover with stock, put on buttered paper, then lid. Cook in slow oven for $1\frac{1}{2}$ hours. To serve, remove lettuces to a colander placed over another pan and press out surplus liquid. Keep lettuce on a hot dish in the oven. Add the liquid to the original sauce, then strain into a small pan reducing contents to 5–6 tablespoons, over a high heat. Remove from cooker, stir in 1 oz of butter divided into small pieces and coat the lettuces with the thick sauce.

MARROWS, SQUASH AND COURGETTES

COURGETTES, FRIED

Use 1 lb of sliced baby marrows, courgettes or zucchini.

Heat 1 tablespoon of oil and one of butter in a small frying pan, add courgettes, replace lid and shake vigorously, leaving on low heat. Shake occasionally and when the vegetables begin to brown at the edges, but are still firm, drain on to a dish. Toss a handful of soft brown breadcrumbs on to the juices in pan, turn up the heat and then crisp, to absorb most of the juices. Return courgettes to pan, stir, season fairly strongly and serve.

AU GRATIN

Slice 2 lbs small marrows into 1 in slices. Put into saucepan with very little water. Season and cook slowly, stirring frequently until all the water has evaporated. Put them into a fireproof dish, add 2 ozs butter, 4 tablespoons cream, 1 beaten egg, and 1 oz grated cheese. Sprinkle with another ounce of cheese and a few knobs of butter. Cook in a brisk oven till the top is slightly brown.

Variations: Chopped ham or frankfurters can be added after the eggs.

STUFFED ZUCCHINI

Scoop out pulp from 4 small zucchini cut in half lengthways. Chop pulp with an onion, clove of garlic, 4 tomatoes, and fry in 1 tablespoon of oil till tender but not browned. Add 4 ozs cooked minced meat. Put back mixture into zucchini skin cases, top with breadcrumbs and grated cheese and cook towards top of moderate oven for 15 minutes. Do not overcook.

STUFFED : (NORTH AFRICAN STYLE)

Use 5 in long courgettes (they are plumper than zucchini). Cut in half and scoop out in same way as previous recipe. Fill with mixture of minced cooked lamb, courgette pulp, partly boiled rice, chopped onion, parsley, a teaspoon of cinnamon, some gravy; season and bind it all with 1 or 2 beaten eggs. The proportions can be varied according to taste.

Stuff half the courgettes, leaving room for the rice to swell. Place on the other halves as lids, and put them in rows in a shallow casserole. Pour on top, but not to submerge, some fresh tomato sauce; cover casserole and bake for an hour in slow oven, or until courgettes are tender and the stuffing well cooked. Add a little water at times if necessary.

COURGETTES BAKED WITH EGGS

Use largish plump courgettes 3–4 ins in diameter, cut into 3 in rounds, and left unpeeled. Boil till they soften a little, but are not fully cooked. Strain, and with a sharp spoon or grapefruit knife scoop out ¾ of the seeds from the top, leaving a base, so they become cups large enough to hold an egg. You may need to remove some of the pulp as well as seeds depending on the size of the eggs and courgettes slices.

Leave them upside down to drain for a few hours, or they can be cooked a day in advance. Give them a final squeeze and pat dry. Break an egg into each marrow 'cup', top with grated cheese and a little cream or butter and bake in brisk oven till eggs are set. A thin slice of raw tomato can be put in the bottom of each 'cup' before the egg, if you wish.

IN RED WINE

Cut 12 baby marrows into ½ in slices and sauté with ¼ lb finely chopped onion in 4 tablespoons olive oil and 2 of butter for 5 minutes, stirring often. Add ¼ pint red wine, salt and freshly ground black pepper. Simmer another 5 minutes. Just before serving, sprinkle with a little lemon juice, finely-chopped parsley and more fresh olive oil if liked.

COURGETTE SOUFFLÉ

Make a purée of small unpeeled marrows, by slicing them, sprinkling with salt, putting in colander with a plate and weight on top for at least an hour, to get rid of their excess moisture. Cook them gently in a heavy saucepan with a ladle of water until they are soft and the water evaporated. Go steady on the water, as the object is to extract it during the cooking and retain the concentrated delicate flavour. Sieve or liquidise them, but with no additional moisture.

Prepare a Béchamel sauce with 1 oz butter, 2 tablespoons flour and just under ¼ pint warm milk (see Page 95). Stir in

½ pint courgette purée, add 1½ to 2 ozs grated cheese and, off the heat, 2 well-beaten egg yolks. Leave to cool then fold in 4 eggs whites (beaten so that they stand in peaks, but not so much that they begin to crumble). Pour into a buttered soufflé dish which the mixture will almost fill, sprinkle the top with grated cheese. Stand in a baking tin filled with water, and cook in the centre of the oven at 355°F. for 40–45 minutes.

Note : Never cook this dish as you would a cheese or other soufflé, because the higher temperature would destroy its texture and flavour. It will rise, but not in such a spectacular way as other soufflés, which is the reason why the dish can be almost filled.

BABY MARROW VARIATIONS

Cut into long thin slices; dip into batter and deep-fry to make fritters. Slice, salt, drain, dry and sauté in oil or butter. Stew in oil with tomatoes, onions and serve hot or cold. Boil, drain and slice while they are still firm, mixed with oil and lemon for a salad.

ONIONS

SOUFFLÉ

To a cup of purée, which should not be watery, add a white sauce made with 1 tablespoon each of butter and flour, and a cup of milk. Cook till thick, add ¼ cup fresh white bread-crumbs, salt and pepper. Beat in 2 egg yolks, then fold in 2 stiffly beaten egg whites. Pour mixture into buttered soufflé or baking dish to two thirds full, place in a pan of water and bake in a moderate oven until firm to the touch – about 30 minutes. It is excellent as a beginning or end to a meal and as an accompaniment for a meat dish, particularly lamb.

STUFFED

Scoop out part of the centres of parboiled large onions, and fill the cavity with a stuffing made of chopped mushrooms, minced left-over poultry, game or meat, flaked cooked fish, chopped nuts or celery – in fact, almost anything you wish. It must be well flavoured and highly seasoned.

AU GRATIN

Boil sliced onions in salted water and drain when cooked. Butter a casserole or pie dish, put in a layer of onions then one of sliced blanched tomatoes, a sprinkle of grated cheese followed by a layer of Béchamel sauce. Repeat until the dish is full, ending with sauce. Sprinkle with breadcrumbs, dot with butter and bake in moderate oven till golden.

ONION SOUP

Cut 4 medium onions into thin rounds and fry in hot butter till pale gold. They must not burn. Add 2 pints good brown stock and simmer for about 20 minutes. Put one thick slice of French bread into each individual bowl, add the soup and sprinkle with cheese. A few crushed cloves of garlic added during the cooking will give a stronger flavour to the onions.

SOUBISE OR ONION SAUCE

For 4 servings, boil 2 medium onions in just enough salted water to cover. Drain, but keep liquid; chop and add to a white or Béchamel (see Page 95) sauce made with 1 oz each butter and flour, $\frac{1}{4}$ pint milk and $\frac{1}{4}$ pint onion water. Season highly. A little cream added last, improves both taste and texture.

À LA MONÉGASQUE

Put 1 lb small onions in a saucepan with a third of a quart of water, 2 wineglasses of wine vinegar, 3 tablespoons oil, 3 tablespoons tomato sauce, 2 ozs of raisins, 2 ozs castor sugar, a bouquet of herbs, a few peppercorns and a little salt. Bring to boil and simmer gently for $1\frac{1}{2}$ hours. Let stand till cold, then chill and serve in a glass dish.

WHOLE IN BÉCHAMEL SAUCE

As for Soubise sauce, but use young whole baby onions. Cook fully before freezing and the dish will only need re-heating.

CHIPS

Cut the blanched parsnips as you would potato chips to thickness of a lead pencil. Dry in a cloth, toss in flour and fry in oil until they are crisp and golden.

CAKES

Boil parsnips and mash thoroughly, working in a little flour, a pinch of mace, pepper and salt, and a large lump of butter. Form into round flat cakes about 1 in thick. Egg and breadcrumb them, and fry in smoking fat till brown. They should be crisp outside the creamy within.

PIE

Boil and drain parsnips and sieve. To about 1 pint of mush add 1 tablespoon of honey, a large dash of ginger and a pinch of spice. Beat in the grated rind and juice of 2 lemons with the yolk of an egg. Line a flan tin with pastry and fill with the mixture. Make lattice crossings with the trimmings from the pastry and bake till golden brown. Pile the beaten egg white (sweetened and flavoured with a little lemon rind) round the edge, and return to the cool oven to set. Serve cold.

FRITTERS

Boil parsnips and chop coarsely. Mix with frying batter and drop a tablespoon at a time of the mixture into deep fat. Cook to a light golden colour.

ALTERNATIVES

Roast, as potatoes round joint of meat. Plain boiled. Made into a purée. Boil and serve in a cream sauce.

PEAS

SOUP, COLD

Cook 1 lb peas in ½ pint chicken stock and a handful of

chopped onion. Sieve or liquidise; add $\frac{1}{2}$ pint creamy milk and chill. Serve with 1 tablespoon of cream, and chopped chives.

Variations: There are dozens for both chilled and hot pea soup with addition of potato, leeks, lemon juice and ham stock instead of chicken.

AU GRATIN

Sauté $\frac{1}{4}$ lb cooked diced ham and a medium finely-chopped onion in butter until onions begin to colour. Combine with 1 lb of cooked peas, $\frac{1}{2}$ pint warm cream. Season, then pour into well-buttered casserole. Sprinkle with 4 ozs of mixed freshly grated Parmesan and Gruyère cheese. Dot with butter and cook under grill till top is golden.

À LA BONNE-FÉMME

Brown 12 small onions in butter with $\frac{1}{4}$ lb of blanched ham or bacon, diced. Remove from pan, stir in 2 teaspoons of flour and, when well blended, $\frac{1}{2}$ pint of stock and bring to the boil. Add $2\frac{1}{2}$ lbs peas, the onions and ham, season, cover and simmer till the sauce has reduced by half and the peas are tender. The time will vary with the size and age of the peas.

À LA BOURGEOISE

Put 1 tablespoon of butter in a heavy saucepan, and when melted, blend in 1 level tablespoon of flour. Add 2 lbs of peas and cook for a few minutes. Cover with boiling water, season, add 4 whole medium onions, 1 chopped lettuce, chopped chives and a little parsley. Cover and simmer until the liquid is reduced by two thirds and the peas are tender. Before serving, add the yolks of 3 eggs, beaten up with a few tablespoons of the hot sauce from the peas. Stir well, but do not let it come to boil again.

À LA FRANÇAISE
See under Lettuce.

DEVILLED NEW POTATOES

Melt 1 tablespoon of butter for each dozen small potatoes in a deep frying pan, add 1 teaspoon made mustard, ½ tablespoon of vinegar, then put in the slightly underboiled potatoes. Brown them quickly and serve very hot.

EN PAPILLOTES

De-frost and dry small new potatoes. Put them on cooking foil or paper, with some leaves of mint, a little salt and 2 ozs butter. Fold the wrapping so potatoes are completely sealed and cook in oven, pre-heated for about 30 minutes.

SALSIFY

PIE

Slice boiled roots in slanting oval slices. Pack into a deep pie dish. Make a white sauce flavoured with enough anchovy sauce to colour it a pale pink. Pour this over the salsify and sprinkle with chopped parsley. Cover with a potato crust and bake as you would a fish pie.

Alternatives: Salsify is also known as the vegetable oyster and, like its fishy namesake, is something to eat by itself with the simplest additions. Its delicate flavour is lost in a composite dish or strong sauce. It is at its best when boiled or steamed; tossed in butter and lemon juice; fried in egg and breadcrumbs, or served with a cream sauce made with white wine or cider.

SCORZONERA

This is the black rooted variety of Salsify, considered to be a finer flavour. It is essential to boil, steam, or bake the roots before peeling them and it must be done while they are still hot.

SEAKALE BEET: SWISS CHARD

Trim the broad mid-ribs free of all green leaf. Cut into pieces, and cook as you would celery.

The green part of the leaf can be cooked and used in the same way as spinach.

SHALLOT

Use as for onions. They have a much better flavour with a subtle suggestion of garlic and are particularly good chopped fine, sautéed in butter until almost burnt and served with steak.

SPINACH BEET

As for seakale beet or spinach.

SPINACH

SOUP

Mix purée of spinach with white sauce, chicken stock, cream, or all three. Serve with croûtons or chopped hard boiled eggs. Possible additions – grated cheese, nutmeg, a dash of lemon juice.

SOUFFLÉ WITH ANCHOVIES

To every $\frac{1}{2}$ lb spinach purée allow the yolks of 3 eggs, 2 ozs grated Parmesan cheese and 3 egg whites beaten to a stiff froth. Mix the yolks with the spinach, stir in the cheese and seasoning and lastly the whites. Pour a thin coating into a buttered soufflé dish, then a scattered layer of anchovy fillets (not packed like sardines). Cover with another layer of soufflé, then anchovies, and end with soufflé mixture, which should only ¾ fill the dish. Sprinkle with a little more cheese. Put in hot oven about 425°F. and when mixture begins to rise, lower heat and cook about 20 minutes at 375°F.

CREPÊS FLORENTINE

Make a thick white sauce with 2 tablespoons each butter and flour. Add 4 tablespoons grated cheese and 4 tablespoons double cream. Season and cook, stirring constantly over low heat until cheese has melted. Add 1 lb cooked, well drained chopped spinach and heat through.

Have ready 12 thin pancakes (the number will depend on size of pan but make more than you think you need). Spread 1 tablespoon of hot spinach mixture on each pancake (which can have been made in advance); roll up and arrange on heated serving dish. Pour over 2 tablespoons melted butter and sprinkle with a mixture of 2 tablespoons grated Parmesan cheese and finely chopped parsley. Suggestion – cooked chopped chicken livers can be added when filling pancakes, or cubes of true paté.

OEUFS FLORENTINE

Allow about 1 lb chopped or sieved spinach to 4–6 eggs. The eggs may be soft or hard boiled or poached till just firm. Arrange a bed of cooked spinach in a buttered oven dish. Lay the eggs on top, making shallow nest for them with the back of a spoon; coat with cheese or mornay sauce and re-heat. Or the dish can be topped with breadcrumbs and grated cheese and browned under the grill.

WITH SULTANAS

Heat 1 lb cooked spinach in hot oil in a wide pan, add a clove of pressed garlic and a little pepper. Keep turning so that it does not fry. When hot put in 1 oz of raisins or sultanas previously soaked for 15 minutes in warm water, and the same amount of pine kernel nuts. Cover the pan and continue cooking gently for another 10–15 minutes.

SWEET CORN

Perfect corn should not be sullied by anything other than melted butter. In this far from perfect world, there is going to be a far greater proportion of also-rans.

These can be used, cut from the cobs, with any assortment of vegetables. They are particularly good tossed in butter with french beans, or added to scrambled eggs and lightly cooked tomatoes.

CORN CAKES
Mince the cooked corn and mix with a thick frying batter. Put a spoonful at a time on a hot griddle or buttered baking sheet, and bake on top, or in the oven till light brown; turn and bake the other side. Serve very hot.

FRITTERS, PLAIN
Dip a tablespoon of cooked corn at a time into frying batter, and fry in deep fat.

FRITTERS, FANCY
Mince 6 ozs cooked corn. Make a batter with 4 ozs plain flour, 2 teaspoons baking powder, ¾ teaspoon salt, ¼ teaspoon paprika, 2 teaspoons castor sugar. Sift these, make a centre well and add 2 egg yolks and 3 tablespoons milk. Mix vigorously with wire whisk gradually bringing in the flour from the sides and adding another 3 tablespoons of milk to make a thick smooth batter. Stir in the corn. Whisk the 2 egg whites till stiff but not dry and fold into batter. Gently float tablespoons of the mixture into deep hot oil so they keep their shape and when puffed up and brown on one side, flip them over and fry till golden. Drain on kitchen paper and serve at once before they lose their crispness. Sprinkle with paprika or a little castor sugar.

128

TOMATOES

SOUP, LIGHT AND PURE

Simmer 1 lb ripe tomatoes with 2 chopped shallots or an onion in 2 ozs butter. Add a bunch of scissored fresh basil, pepper, salt, and a large teaspoon of sugar. Just cover with water and cook gently for 20 minutes. Strain to get rid of pips and skin, return to pan and gradually add thin cream or rich milk, stirring till hot but not boiling. Serve with more basil on top.

SWEDISH STYLE

Melt 2 tablespoons dripping in large pot and sauté 2 chopped onions and clove of garlic for 5 minutes. Add 2½ lbs ripe tomatoes and 2 sticks of celery cut up. Pour in 1 pint chicken stock, cover and simmer for 30–40 minutes. Sieve. Melt 2 ozs butter in the pot and stir in 2 tablespoons of flour, then the strained tomato. Stir till well blended, then add 2 tablespoons of dry sherry. Thin with chicken stock or cream if necessary and serve sprinkled with parsley.

TOMATO SOUP WITH ORANGE

Slice an onion and a carrot finely and put in large pan with 2 lbs ripe tomatoes, a bay leaf, strip of lemon rind, 6 peppercorns, 2 pints of stock. Add salt and simmer for an hour with lid half off. Sieve. Melt 1½ ozs flour and stir in 1½ ozs butter, mix well then dilute with the tomato pulp. Grate the skin of an orange finely so that the pith is not reached. Rub a lump of sugar on to the remaining skin to collect the zest and add these to the sauce with more sugar to taste. Simmer 5 minutes or more till the soup is concentrated. Add a gill of cream at the last moment and garnish with grated orange rind.

STUFFED

Cut tomatoes in half or slice off top to use as a lid. Scoop out the seeds and some of the pulp. Leave upside down to drain while preparing the stuffing. Boil 4 ozs long grain rice till tender. Fry a small chopped onion in 1 tablespoon of oil for a few minutes, add 2 ozs finely chopped mushrooms and cook 5 minutes more. Mix in 8 ozs cooked diced or minced beef, ham or chicken, 2 tablespoons of chopped parsley, or 1 tablespoon of scissored basil, and mix in some of the tomato pulp. Season the tomato cases and the stuffing. Fill them, put on the lids, unless they are halves, and bake for about 15 minutes.

Variations: Breadcrumbs (instead of rice), grated cheese, chopped hardboiled eggs, shredded almonds, pine nuts, buttered leaf spinach, small pieces of crisp fried bacon.

STEW

Fry 3 chopped small onions in a tablespoon of oil till golden, add 3 rashers of lean chopped bacon. Cut 8 small tomatoes into quarters, add to pan and cook gently for 5 minutes. Slowly stir in 1 tablespoon of flour. Add 2 tablespoons white wine a little at a time, then 8 stoned and chopped black olives. Simmer for a few minutes to cook the flour and at the last moment blend in 1 oz of butter.

PURÉE

Cook a small sliced onion in oil until soft. Add 1 lb cut-up tomatoes, a sprig of thyme, small bunches of parsley and basil, a bayleaf, teaspoon of sugar, salt and pepper. Stir well with a wooden spoon, crushing the tomatoes to extract the juice. Simmer gently until they are reduced to a pulp, and sieve.

TURNIPS

GLAZED

Boil small young ones, till not quite done. Drain. In the pan melt enough butter with an equal quantity of soft brown sugar, to coat them well. Put back turnips, put on lid, and cook on fairly high heat, shaking all the time until the turnips are well brown and sizzling. Very good with duck.

Alternatives: Turnips can be boiled and mashed with butter or bacon fat, make into a soufflé or cooked in any way you would use carrots.

MIXED VEGETABLES DISHES

VEGETABLE PIE
Put into a fireproof dish $\frac{1}{2}$ lb cooked tomatoes, $\frac{1}{4}$ lb of mushrooms lightly fried with a dozen spring onions. Stir 2 ozs semolina into a breakfast cup of milk on low heat for a few minutes till thick. Take off heat and stir in 2 ozs grated cheese, seasoning and nutmeg. Spread on vegetables, top with grated cheese, breadcrumbs and dots of butter. Bake in moderate oven for 25 minutes.

COURGETTES WITH TOMATOES
Use alternate layers of blanched, sliced courgettes, and peeled sliced tomatoes. Between each layer add onions sautéed in oil with a little garlic, and sprinkle grated cheese, finishing with tomatoes and cheese. Top with a few browned crumbs. Pour in a little stock to moisten and bake in a moderate oven for about an hour.

HOME GROWN RATATOUILLE
Use a wide, heavy pan, with well-fitting lid. Heat 2 tablespoons of oil and gently fry small onions or 2 large sliced ones with 2 cloves of crushed garlic. Add 6 thickly sliced courgettes, 1 lb peeled and halved tomatoes, $\frac{1}{2}$ lb of blanched French beans, 2 bay leaves, sprig of thyme, tablespoon of scissored basil, seasoning. Put on the lid and simmer very gently until the vegetables are tender, but still separate and not turned into a mush. Serve sprinkled with parsley, slightly warm or cold, NOT chilled.

MIXED VEGETABLES WITH BUTTER
Mix the following cooked vegetables, sliced, whole or cubed, in butter for a few minutes, season and sprinkle with any favourite fresh herbs – broad beans, French beans, peas, new

131

potatoes, baby turnips, white or golden beet, etc. Pickling onions browned in butter can also be included.

Variations: A cream sauce may be used instead of butter.

ARTICHOKE HEARTS WITH ASPARAGUS AND MUSHROOMS
6 cooked artichoke hearts (bottoms) cut in quarters, $\frac{1}{4}$ cup cooked button mushrooms, 18 cooked asparagus tips, 2 cups Mornay sauce (as for Béchamel with the addition of 3 tablespoons each of thick cream and Parmesan cheese), $\frac{1}{2}$ cup Gruyère cheese. Put mushrooms and sauce into a shallow oven-proof dish, arrange asparagus in centre and artichokes round the edge. Sprinkle with cheese and a little melted butter. Put under grill or into very hot oven until the cheese is light gold.

RECIPES : HERBS

Their principle uses have been suggested in a previous chapter, together with the foods with which their individual flavours are most in harmony. Some, though, can be used as complete dishes on their own, or as accompanying sauces.

BASIL SOUP
Melt a large handful of sliced basil leaves gently in butter. Add them to potato purée, thinned with milk or mild stock. The proportions and textures are a matter of individual choice.

BASIL OMELETTE, SWEET
Pound 2 large leaves of sweet basil with 1 dessertspoon of sugar into a fine pale green powder. Separate 2 eggs, and beat the yolks lightly with another tablespoon of sugar, then gently fold in the stiffly beaten whites.

Heat a small nob of butter in an omelette pan, pour in the egg mixture and cook gently until the omelette is just setting, with the top still moist and fluffy. Slide on to a warm plate without folding, sprinkle with the basil and sugar and a few drops of lemon juice.

BUTTERS
Cream 4 ozs butter; drop by drop beat in 1 tablespoon lemon

juice, then beat in 2–3 tablespoons of freshly chopped parsley; mixed green herbs, or tarragon. Chill and serve at last moment on grilled meats or fish, or use to enrich soup or sauce.

DILL SAUCE

Make a basic white sauce using cornflour; cook a few minutes, then beat in, off the heat, an egg yolk, a little cream, and 2 tablespoons of chopped dill. Re-heat gently, but not to boiling point. Excellent with fish, particularly eel.

DILL VINAIGRETTE

Beat an egg yolk with 4 tablespoons cream or sour cream, then gradually beat in $\frac{1}{4}$ pint French dressing, as though making a mayonnaise. Season to taste with lemon juice and stir in 2 tablespoons of chopped dill. Alternative herbs are parsley, chives, tarragon, chervil. To use with cold eggs, vegetables, and cold or hot fish.

FENNEL SAUCE

Add 1 or 2 teaspoons of chopped fennel to $\frac{1}{2}$ pint Béchamel sauce a few minutes before serving. Stir well but do not allow it to boil or the fresh flavour of the herb will be destroyed. Use with fish.

'HERB' SOUP

Heat 1 tablespoon of butter in a saucepan. Add $\frac{1}{4}$ lb of sorrel, $\frac{1}{2}$ a small lettuce, and 1 tablespoon of chervil, all chopped. Simmer gently for 30 minutes without browning, add $2\frac{1}{2}$ pints of stock and simmer another 30 minutes. Ten minutes before serving, mix 2 egg yolks in a bowl with a little warm, but not hot, stock. Take the saucepan off the heat, slowly stir in the yolks; replace the pan. Stir until the soup thickens slightly and just before serving add $1\frac{1}{2}$ ozs butter divided into small pieces. Serve with croûtous of fried bread.

MINT SAUCE

Mix 3 tablespoons of finely chopped mint with $1\frac{1}{2}$ of castor

133

sugar and 3 of wine vinegar. A little boiling water can be added to help dissolve the sugar in the mint, before adding the vinegar.

MINT JELLY

Stir 8 ozs sugar and ½ pint of water over steady heat until sugar has dissolved. Blend ¼ pint white malt vinegar with 1½ level dessertspoons of powdered gelatine; blend with sugared water and continue stirring until the gelantine dissolves. Cool, then stir in 3 tablespoons finely chopped mint. Leave on heat and stir occasionally as the mixture thickens, to make sure the mint is evenly distributed. When the jelly begins to stiffen, pour it into jars and seal.

MINT ICE

As first course, or as a refresher after meat course. Make a syrup with ½ lb sugar and 1 pint of water with a generous dash of lemon juice and 2 tablespoons of chopped mint. Strain and freeze. Half way through freezing, add the beaten whites of 2 eggs and re-freeze.

PARSLEY SAUCE

To a pint of Béchamel or cream sauce (see page 95) add 1 helped tablespoon finely chopped parsley and 1 teaspoon of lemon juice. Simmer for a few minutes so that the sauce is well flavoured and the parsley looses its raw taste.

Variations: For boiled fowl or veal make a white sauce with 1 oz each of butter and flour and gradually add ½ pint of the liquid from the fowl or meat until it thickens. More liquid can be added if wished, or hot milk or cream. Then add the parsley. For fish dishes, use fish stock in place of milk.

SORREL SOUP : (CREAM) (1)

Soften 2 ozs chopped onion in 1½ ozs butter; stir in 1½ ozs flour, blend in 1¾ pints warm milk. Stir till boiling and simmer for 5 minutes. Then beat in 3 tablespoons of sorrel purée (made from boiled, drained and sieved or pulverised sorrell leaves), add 1 gill of cream, adjust seasoning and add lemon juice and sugar to taste.

Note: If sorrel is too sharp for your liking, add equal quantities of spinach and sorrel.

SORREL SOUP (2)

Make a potato purée by boiling together 2 sliced leeks or medium onions, 8–10 ozs of peeled, diced potatoes, in 1¾ pints of stock, plus a bay leaf, for 40 minutes. Then liquidise. Add to this as much cooked sorrel as you like and re-heat. Cream can also be added during the final heating.

SORREL SAUCE

Melt ½ oz butter in saucepan, add ½ oz flour, and cook gently until straw-coloured. Add 3 tablespoons of sorrel purée, but do not stir. Cover and leave in a warm place for 10–15 minutes. Then add a squeeze of lemon juice, seasoning, ½ teaspoon sugar, 1–1½ gills veal stock or water, ½ teaspoon Meat glaze or Marmite. Simmer for 5 minutes until it has the consistency of cream. Adjust seasoning and sugar, then add 2 tablespoons of sour cream. Particularly good with veal or fish.

RECIPES: FRUIT

ICE CREAMS

Quickest and best. Stir ½ pint of any fruit purée and 1½ tablespoons castor or icing sugar into ¾ pint whipped cream. Put into containers and freeze quickly, so that no particles form.

WATER ICE

Basic method. Mix equal quantities of fruit purée with light or medium syrup and freeze. The juice of 1 orange or lemon can be added. Light syrup is 2 cups sugar dissolved in 4 cups of water. Medium is 3 cups sugar to 4 of water.

BLACK CURRANT SORBET

Boil 4 ozs sugar with $\frac{1}{2}$ pint of water gently for 10 minutes.
Let it cool. Meanwhile stew $\frac{1}{2}$ lb of black currants in the
minimum of water for 10 minutes. Sieve. Make pulp up to
1 pint with water. Cool, mix with syrup and 1 teaspoon of
lemon juice. Pour into ice tray and freeze till almost set.
Whisk 2 egg whites until just stiff but not dry. Tip the frozen
fruit into a chilled bowl, break it up with a fork, and fold in
the egg whites, then re-freeze.

GOOSEBERRY SORBET

Simmer 1 lb hard green gooseberries and the grated rind of
a lemon in $\frac{1}{2}$ pint of water to the consistency of a soft pulp
– no need to top and tail. Sieve, then add the juice of the
lemon and 3 ozs of sugar. (1 level teaspoon of powdered
gelatine added now will prevent the sorbet from becoming
'chippy', but is not essential.) Freeze for about 30 minutes.
Whisk 2 egg whites till stiff and stir in the half-frozen fruit
mixture. Mix gently and re-freeze.
Note: Sorbets look particularly attractive frozen in halved,
scooped out fruit cases, such as lemon, orange and grapefruit.

This recipe is suitable for other fruits, notably apricots and
red currants, raspberries and strawberries. The last two need
less sugar and can be used raw.

BLACKBERRY ICE CREAM

Boil 6 ozs sugar with 1 gill of water and the leaf of either a
sweet geranium, a few black currant leaves or an elderflower.
Leave to cool then strain and stir into the pulp from 1 lb of
sieved, cooked or raw blackberries. Slightly whip $1\frac{1}{2}$ gills of
cream and fold into the blackberry mixture. Freeze.

APRICOT ICE CREAM

Sprinkle 1 lb apricots with 3 ozs castor sugar, and bake with-
out water in a covered casserole in a gentle oven till soft.
Sieve and add 1 tablespoon lemon juice and 1 tablespoon
Cointreau. Whip $1\frac{1}{2}$ gills double cream lightly and fold in

the apricot purée. Partly freeze and when the mixture reaches a state of slush, stir in a few blanched sliced almonds.

RUSSIAN RASPBERRY PUDDING
Put 1 lb raspberries into a shallow fireproof dish, sprinkle with 2 tablespoons of castor sugar, and put into the centre of a slow oven until the raspberries are hot right through. Beat up $\frac{1}{2}$ pint of sour cream with 2 eggs, 1 tablespoon of flour and 1 tablespoon of castor sugar. Pour the mixture over the raspberries and put back in the oven at the same heat (Mark 2, 300°F.) but nearer the top. Cook about 45 minutes till the top is golden and firm. Sprinkle with a little more sugar before serving. Eat hot or cold, but more exotic in its hot state.

SUMMER PUDDING
This is a dish which is unlikely to turn out the same twice in a lifetime, as there are no rigid rules or quantities. Line a pudding or soufflé dish with thinly cut slices of crustless bread, overlapping and filling any gaps with smaller pieces, so no juices can escape. Fill with any mixture of hot, sweetened fruit, being careful not to add too much juice. The fruit should only be heated to the point where the sugar dissolves, and not stewed, otherwise the individual flavours will be lost. The usual ingredients are currants and berries of all kinds, such as white currants and raspberries. Some can be left raw as a bite and slight contrast. But there is no reason why cherries and other stone fruit should not be included. It is as adaptable and mysterious as a cottage pie.

Keep back some of the liquid and when the dish is full, cover with more bread slices. Put a plate or saucer on top which fits exactly inside, then weight it heavily. Leave in a refrigerator for 24 hours. Serve on a large dish with the extra juice spooned on to any piece of bread that is not completely soaked through. Cut it into wedges and serve with softly whipped cream or clotted cream.

PEARS BAKED WITH HONEY AND CIDER VINEGAR
A way to use small iron-hard cooking pears. Peel 2 lbs pears,

137

leaving them whole and with the stalk attached. Put them, stalk upwards, in an earthenware pot, and add ½ gill each of cider vinegar and honey, 4 tablespoons of white sugar, 1 gill (¼ pint) of water and a strip of orange peel. Cover the pot and put it at the bottom of the oven, No. 1–3, 290° to 330°F.

It is not possible to give exact timings as the fruit varies so much – as well as ovens – they can be anything from 2–6 hours. The pears must be still whole, but quite tender, with the juice a dark amber colour. They are delicious cold alone, or with whipped cream, or served with cold duck, ham, pork, or anything which is in sympathy with a sweet-sour taste.

COLD FRUIT SOUFFLÉS

Beat 3 egg yolks with 2 ozs sugar over hot water till thick. Cool slightly and add 1 level dessertspoon of powdered gelatine dissolved in 2½ tablespoons of hot water. Stir in ¼ pint of fruit purée, a tablespoon of sherry or maraschino, and when quite cold, ¼ pint lightly whipped double cream. Let it stiffen very slightly, then fold in 3 stiffly beaten egg whites. Freeze. Serve with split, toasted almonds.

HOT APPLE SOUP

Use the sieved pulp from 1 lb sour cooking apples. Add it to a saucepan with 4 pints of water, the grated rind of a lemon, 2 tablespoons of sugar, 8 ozs each of sultanas and currants and simmer gently for 30 minutes. Meanwhile, melt 1 tablespoon of butter in a small thick saucepan. Slowly stir in 1–2 tablespoons of cornflour and gradually add this to the soup, stirring all the time. Simmer for another 10 minutes then serve. You can be more generous with the sugar and butter if it suits your taste.

CHERRY SAUCE

To serve with duck, poultry or game.

Cook 8 ozs cherries gently in 1 gill of water till soft. Remove stones and sieve. Put back into juice and add 2 ozs seedless raisins, 1 teaspoon sugar, seasoning, and ½ pint of chicken stock. Blend 2 ozs of butter with 4 teaspoons of flour, dilute with a little of the liquid and add to the cherry sauce. Mix well, bring to the boil and simmer for 10 minutes, stirring frequently.

CHERRY DUMPLINGS

Cover 1 lb of cherries with 4 ozs of sugar and leave them for several hours. Into 8 ozs of flour blend 2 eggs, a pinch of salt and 2 tablespoons of water. Knead into a stiff dough and leave for 1-2 hours. Then roll out pastry very thinly and cut into circles.

Strain the juice from the cherries and set aside. Stone cherries and cook them with 1 cup of water for 5 minutes. Put a few cherries into each circle of pastry; fold, press the edges together and put into a large pan of boiling water to cook for 10 minutes. Use both juices from the cherries, mix with 1 gill of sour cream to make a sauce for the dumplings.

FRUIT FRITTERS

Make a thick coating batter with 4 ozs of flour, 3 eggs and 2 extra egg white, with 1 tablespoon of melted butter. Coat fruit generously in batter, fry in butter or cooking fat until golden and crisp, then serve sprinkled with sugar and cinnamon. Suitable fruits are large ripe plums, small sweet apples (with stems left on), cherries, white or black currants. Or a mixture of fruits can be used and served, piled in a large dish.

RHUBARB AND GINGER COMPÔTE

Boil 1 cup of sugar and 1 cup of water rapidly for 5 minutes, add 2 tablespoons syrup from jar of preserved ginger, or ½ teaspoon of powdered ginger. Put in 1-1½ lbs of sliced rhubarb. Cover and poach gently 5-7 minutes. Serve chilled.

GOOSEBERRY TANSY

Cook a pint of gooseberries with a large lump of butter in a thick saucepan, covered, till they are quite soft. Beat 2 eggs in a basin, add a handful of fresh white breadcrumbs and ½ a cup of sugar. Blend this into the gooseberry pulp over a low heat, stirring very gently until the mixture is cooked firm to the consistency of a solid omelette. It will curdle if the heat is too high. Put into a hot dish, sprinkle with coarse sugar and serve with hot cider, melted apple, or blackberry jelly.

The trick is to use only enough egg and breadcrumb mixture to absorb the buttery liquid from the cooked fruit, and make it all 'bind'. You may have to add more butter if the fruit is very dry, or more crumbs if the gooseberries are very ripe.

Raspberries, any of the hybrid berries and strawberries all make delicious tansies.

DAMSON CHEESE

Put damsons in a large earthenware dish in the oven and cook slowly until the juice runs freely and the fruit is soft. Stir well and rub through a sieve. Crack the stones and add the kernels to the pulp to give it a distinctive almond flavour. Put the pulp into a fireproof pan, adding 1 lb of sugar to every 1 lb of pulp (the sugar should have been warmed in the oven). Boil well until it jellies, then put the cheese into straightsided preserving jars. It should not be used for 6 months and improves up to 2 years. It is at its best when it has shrunk a little from the sides of the jar and the top is just starting to crust with sugar. Serve it as a dessert turned into a dish and pour port wine over it.

Note: This dish is not intended to be deep frozen, as the cheese needs to mature gradually. It is intended for using whole frozen damsons, or pulp that you have not had time to deal with during the summer harvesting.

RECIPES : RANDOM HARVEST

See also: TREATMENT (Page 71)

BILBERRIES

They make exceptionally good fillings for tarts and pies and summer puddings. They can be turned into fool, mousse, cheese (see damsons) or used raw and unfrosted as you would any of the berries in the fruit section.

BLACKBERRIES

See fruit section.

COCKLES

However they are to be used, prepare them first in the way described for Treatment of Random Harvest, (Page 71), which will leave you with the fat, succulent, barely cooked cockles and pure strained cockle juice. If they are to be used in a hot dish, never let them cook long or fast; they will turn leathery. They are at their best simply de-frosted, their juice heated, thickened with a little cornflour, then cooled and poured over them.

COCKLE SAUCE

Soften a large sliced onion in $1\frac{1}{2}$ ozs of butter. Remove pan from heat and stir in enough flour to absorb the butter. Add $\frac{1}{4}$ pint of milk and $\frac{1}{2}$ pint of cockle juice, bring to simmer and cook over very low heat for about 15 minutes, stirring often. Beat 1 egg yolk with 1 dessertspoon of lemon juice, add a tablespoon of sauce from the pan and stir this mixture back into the sauce, adding cockles at the same time. Heat through without boiling. Add 2 dessertspoons of parsley or chives, chopped, and seasoning. Cockle juice will be salty, so beware! A little cream added at the last moment makes this sauce even more delicious.

It is excellent served with haddock, cod, turbot and halibut. On some parts of the coast – notably Norfolk – it is served with roast chicken, a strange association which works out most happily.

COCKLE SOUP

Using 2 quarts of cockles.

Heat 2 ozs butter and fry 2 sticks of finely chopped celery and 1 finely chopped medium sized onion. Off heat, blend in 2 ozs of flour, return to heat and gradually add 1 pint of milk. Stir till simmering then add 1½ pints cockle juice, 1 tablespoon chopped parsley and seasoning. Simmer gently for 15 minutes. The cockles can be added at any stage during the simmering, depending on how well you like them cooked, but they should not actually boil. Stir in 2 tablespoons of cream just before serving. This amount makes enough for 6 people.

COCKLE PIE

Make a creamy sauce (described above), butter a pie dish and sprinkle with brown bread crumbs. Put in a layer of cockles, cover with a layer of sauce then a layer of crumbs (these may be toasted), then more butter, more cockles and more sauce until the dish is full. Cover with a thin crust of mashed potatoes and bake in a quick oven, Gas 5, 400°F. for about 20 minutes until crisp and golden on top, but the cockles not at boiling point. Serve with a squeeze of lemon juice.

Alternatives: Grated cheese may be added to the breadcrumb layers, and the topping for the pie can be of pastry rather than potatoes.

Note: Cockle juice can be reduced by quick boiling if too watery, or thinned with milk or water if too intense.

FRIED COCKLES

These in many remote coastal areas are esteemed more than oysters. But in the initial preparation (Treatment Page 71) they must be plucked from the pan the moment the shells open so that they are almost raw.

They can be rolled in flour and pepper and fried in bacon fat till very brown and crisp; coated with flour, egg and breadcrumbs; dipped individually into batter and deep fried, or made into pancakes, cooked on both sides, sprinkled with lemon juice and rolled up. They should be served at once.

OTHER USES

For most of the recipes in which shrimps are normally used, such as Vol au Vent, potted or cocktail.

ELDERFLOWER

ELDER AND LEMON SORBET

In a saucepan, clean beyond reproach, put 1 pint of water, 6 ozs of sugar and the thinly pared rind (no pith) of 3 lemons and 3–4 heads of elderflower. Dissolve completely, then boil rapidly for 6 minutes. Set aside to cool, add the juice from the lemons, strain and freeze. Stiffly whipped egg white may be added to give a more frothy texture.

ELDERFLOWER FRITTERS

Dip elderflowers into a coating batter and fry quickly in oil. Sprinkle with sugar to serve.

ELDERBERRY SAUCE

Cook 2 pints of ripe berries, picked from the stalk, slowly in a covered dish, with ½ pint of wine vinegar. This can be done in the oven and left there to cool. Leave for a day, then strain the juice into an enamel pan. Add 1 large minced shallot, a piece of bruised root ginger, 1 teaspoon of cloves and a few peppercorns. Sugar may be added to taste. Boil till the sauce reduces and thickens, then strain, cool and freeze, or store in jars.

ELDERBERRY ROB

This makes hot drinks for winter colds. Put a dish of elder-berries into a hot oven until the juice runs; strain off, and to 1 pint of juice add ½ lb of sugar and ¼ teaspoon (or large broken stick) of cinnamon; cover and boil slowly till thick.

HORSE MUSHROOMS, STUFFED
(Psalliota arvenis)

These are a round, less delicate version of the field mushroom, which make excellent 'cups' for stuffing and baking.

WITH TOMATOES

Pepper and salt the mushrooms inside. Don't peel. Add a dab of bacon fat or butter. Press in a neatly fitting skinned tomato. Turn them upside down into an oven dish, put thin strips of bacon fat over their domes and bake in a fairly hot oven until the mushrooms are cooked and the tomatoes have made a rich sauce.

The thickest and worst shaped horse mushrooms should be baked on their bottoms rather than their heads. Slice a little from the cap so they stand firm, pack them as before, and top with bacon strips. Wedge them tightly together and pour a little stock into the pan .

Alternatives: Fill the mushroom cups with chopped ham and a raw egg, topped with grated cheese, cream and butter; or with savoury rice or any of the stuffings suitable for vegetable marrow.

PUFF-BALLS (LYCOPERDON)

These come in small, medium and giant sizes as big as a football. All have the same texture – smooth, solid and white. All must be used young. When they have started to go soft and puffy they are useless. A small puff-ball will not grow into a medium one, nor will a medium develop into a giant; so size is no indication of age. You judge this by the firmness of the flesh.

Walnut Size: Stew gently in milk until tender; strain and make a white sauce with the milk. Put back the puff-balls, season, and serve on hot toast.

Alternatively, dip into batter, deep fry, and serve as a necklace to a purée of spinach.

Hens' Egg Size: Wipe, roll in seasoned flour and put in a saucepan with milk not quite covering them. Add a bay leaf

and a little chopped onion; simmer gently till soft. Lift them
on to a hot dish. Thicken the milk with butter and cornflour,
season with a light hand, cook thoroughly and pour back
over the puffs. This is a very white dish, calling for a
spectacular garnish, such as parsley and a few cranberries, or
bilberries.

Giants: These have the texture and flavour of sweetbreads.
They are too large to cook whole. Slice into rounds $\frac{1}{2}$ in
thick. Dip the slices into well-beaten egg and milk, then fine
breadcrumbs. Press the crumbs in firmly and put in the
refrigerator for an hour or more to become settled. Fry in
bacon fat on both sides, drain and serve very hot, sprinkled
with cider or vinegar. If this sounds too much like chip
shop treatment, use any piquant sauce.
Note: Cubes of the giant puff-ball do remarkable things
when added to steak and kidney pie and pudding, or to
casseroles.

FAIRY RING : (Champigone Marasmius Oreades)
Delicate little things of which you can use both stem and
cap. A slightly sweet taste with a smell of almonds. They
can be tossed in salads or added liberally to stews. They go
particularly with rabbit dishes.

MUSSELS

MOULES MARINIÈRE
There are uncountable variations for this classic dish. Here
is a basic one. Simmer together 2 chopped shallots, 1 glass
of dry white wine and 1 of water, with 2 ozs of butter mashed
with 1 dessertspoon of flour. Add 4 pints of mussels and cook
on a high flame, shaking the pan often, so the mussels at the
bottom will not be overcooked and those at the top raw. As
soon as the shells open, strain the liquor. Remove one shell
from each mussel and put the rest in a large bowl to keep

hot. Add 2 ozs butter to the liquid, 2 tablespoons of chopped parsley; then re-heat and pour over the mussels.

Alternative: The butter and flour paste can be omitted and the strained liquid thickened with a roux while the mussels are being kept warm.

MUSSEL PIE

Put shell-less mussels into a buttered oven dish and sprinkle with 1 tablespoon of white wine. Scatter with finely chopped parsley and 2 small onions, also finely chopped, season with pepper and salt, and cover flat with white bread crumbs. Dot with butter and fill the dish with mussel liquid, just barely to the crumbs, so they are damp but not covered. Bake in a moderate oven till the top is crisp and brown and the fish hot and buttery.

MUSSELS WITH RISOTTO

Melt 2 ozs of butter and the same quantity of oil in a deep pan. Add 2 chopped onions and cook until lightly browned. Add 1 lb rice; stir well and cook for 15 minutes; then add 1½ pints of mussel liquor, 4 tablespoons of white wine, a pinch of saffron, salt and pepper, and mix thoroughly. Simmer gently for about 25 minutes, stirring occasionally. Add the mussels 10 minutes before serving, and sprinkle with 2 ozs of grated Parmesan cheese and 2 ozs of melted butter.

MUSSELS IN GARLIC CREAM

If possible, use very small mussels. Reduce some of their cooking liquid, thicken with a small quantity of roux, add cream and crushed garlic, stirring all the time over a low heat till the flour in the roux is cooked, and the sauce creamy. Half-fill heated soup bowls with mussels and pour the hot cream mixture over them. The dish is a cross between a sauce and a soup, with a greater proportion of mussels to liquid.

MOULES ROUCHELLAISE

Make a sauce of butter, cream, fine chopped shallots, garlic, a little curry powder and parsley. Serve with pre-cooked mussels.

Alternatives: Mussels can be used in any of the ways described for cockles.

SAMPHIRE: *See* Random Harvest Treatment. (Page 73)

TYPES OF FREEZER

There are two basic kinds of freezer, the dive-in head first chest type, long and low like a sideboard with a top opening lid, and the reach-in upright cabinet type with a front opening door.

CHEST FREEZER
Advantages: Cheaper to run and buy because it loses less cold air each time the lid is opened. It can store a large quantity of bulky food.

Disadvantages: They take up a good deal of floor space. You can lose things in them unless you have a careful system of fitted baskets and easily identifiable containers. You need to keep an up-to-date chart of the contents, or risk your memory and grope around, head down until you find what you are looking for.

UPRIGHT FRONT-OPENING FREEZER
Advantages: They use less floor space and are easier to load and unload. They have more inside fittings like shelves, drawers and baskets, so that food is easier to find.

Disadvantages: The front-opening door causes a greater loss of cold air than the chest types, because cold air moves downwards and will 'fall out' of an upright freezer if the door is left open longer than necessary. They have to be defrosted more often than chest types but it is easier to do. They cannot take such large or as bulky quantities as the chest freezers.

WHERE TO PUT THE FREEZER

It is something you will use for long term storage rather than as a larder, so there is no need for it to be in the kitchen. It can be in a spare bedroom, passage, shed, garage, or adjoined to the house in its own specially built quarters.

MAIN REQUIREMENTS

The chosen place should be cool, dry, well ventilated, with a suitable electric socket to take an earthed plug. If there is any risk of dampness, protect the freezer base from rust by raising it from the floor so that air can circulate beneath. Perfectionists underseal them as you would a car. If kept in an outhouse or beyond constant supervision, keep the freezer or the shed locked – BOTH if you are away for more than a day.

EMERGENCIES

Don't panic if there is a power cut or the freezer breaks down. Keep the lid or door firmly shut and resist the urge to peep in and see how things are going. A large, well-filled freezer will keep the contents frozen for at least 12 hours, as long as you let in no warm air.

Power cuts seldom last that long and breakdowns are rare. If this awful calamity does happen, check that the wiring, plug and fuses are in order before telephoning the freezer manufacturer's local agent for service.

Keep the name and phone number of the service agent stuck to the side of the freezer.

If the serviceman needs the food to be removed, wrap it in several layers of newspapers and then into blankets and keep on a cold floor.

If the repair is going to take some time, the service depot can usually lend a replacement freezer while the work is done.

If food has started to thaw out, it is still safe to re-freeze it as long as there are ice crystals still present. Once these have melted, you must have a care. The best way to rescue thawed-out produce is to cook them at once and re-freeze. Anything which has thawed so thoroughly that it no longer feels cold, must be thrown away.

INSURANCE

The contents of your freezer can be insured, which makes the loss of hard-earned home grown produce easier to bear. Losses are more often due to carelessness than an electrical or mechanical fault in the freezer. Someone accidentally turns off the switch, or borrows the freezer socket to use another implement, and forgets to plug in again. Watch the small print of any insurance agreement to make sure you are covered for such incidents.

MOVING HOUSE

Ask the removers if they can handle the freezer when loaded. If the move takes place within a day the food will not suffer if it is the last item on the van the first off the other end with a plug ready to switch on. If the freezer has to be unloaded, pack the food into tea chests with 'dry ice' or swathed in newspapers and blankets.

GENERAL RECIPES

INDEX OF RECIPES